London Mathematical Society Student Texts 109

ADE

Patterns in Mathematics

PETER J. CAMERON
University of St Andrews

PIERRE-PHILIPPE DECHANT
University of Leeds and University of York

YANG-HUI HE
*London Institute for Mathematical Sciences, and
Merton College, University of Oxford*

JOHN MCKAY
Concordia University, Montréal

CAMBRIDGE UNIVERSITY PRESS

Shaftesbury Road, Cambridge CB2 8EA, United Kingdom

One Liberty Plaza, 20th Floor, New York, NY 10006, USA

477 Williamstown Road, Port Melbourne, VIC 3207, Australia

314–321, 3rd Floor, Plot 3, Splendor Forum, Jasola District Centre,
New Delhi – 110025, India

103 Penang Road, #05–06/07, Visioncrest Commercial, Singapore 238467

Cambridge University Press is part of Cambridge University Press & Assessment,
a department of the University of Cambridge.

We share the University's mission to contribute to society through the pursuit of
education, learning and research at the highest international levels of excellence.

www.cambridge.org
Information on this title: www.cambridge.org/9781009335966
DOI: 10.1017/9781009335935

When citing this work, please include a reference to the DOI 10.1017/9781009335935

First published 2025

A catalogue record for this publication is available from the British Library

Library of Congress Cataloging-in-Publication Data
Names: Cameron, Peter J. (Peter Jephson), 1947– author. | Dechant, Pierre-Philippe, author. | He,
Yang-Hui, 1975– author. | McKay, John, 1939-2022, author.
Title: ADE : patterns in mathematics / Peter J. Cameron, Pierre-Philippe Dechant,
Yang-Hui He, John McKay.
Description: Cambridge ; New York, NY : Cambridge University Press, 2025. |
Series: London Mathematical Society student texts ; 109 | Includes bibliographical
references and index.
Identifiers: LCCN 2025007650 (print) | LCCN 2025007651 (ebook) |
ISBN 9781009335966 (hardback) | ISBN 9781009335980 (paperback) |
ISBN 9781009335935 (epub)
Subjects: LCSH: Dynkin diagrams. | Representations of Lie algebras. | Representations of Lie groups.
Classification: LCC QA252.3 .C358 2025 (print) | LCC QA252.3 (ebook) |
DDC 512/.482–dc23/eng/20250515
LC record available at https://lccn.loc.gov/2025007650
LC ebook record available at https://lccn.loc.gov/2025007651

ISBN 978-1-009-33596-6 Hardback
ISBN 978-1-009-33598-0 Paperback

Cambridge University Press & Assessment has no responsibility for the persistence
or accuracy of URLs for external or third-party internet websites referred to in this
publication and does not guarantee that any content on such websites is, or will
remain, accurate or appropriate.

For EU product safety concerns, contact us at Calle de José Abascal, 56, 1°, 28003 Madrid,
Spain, or email eugpsr@cambridge.org

To John McKay, with reverence and fond memories.

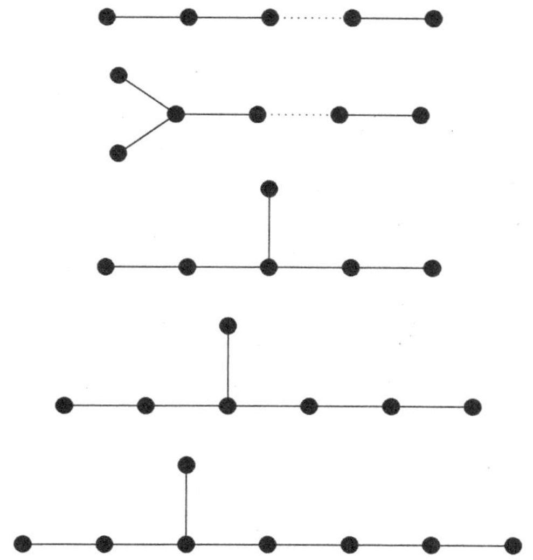

Contents

Contents

Preface

When we were first introduced to mathematics as children, we saw it as a whole – a world of patterns, shapes and numbers – and marvelled, albeit at a puerile level, at this wonderful tapestry that is mathematics. As we grow older and delve more deeply into the matter, we slowly learn of the distinct regions into which this tapestry is partitioned: algebra, geometry, combinatorics, ...

The price of this maturation, however, is that while we can still examine the beauties of an individual problem, the ability to appreciate the interconnectivity between problems slowly fades. This is, of course, inevitable; the vast growth of the corpus of knowledge so overpowers the human intellect that compartmentalisation of expertise is a natural consequence. By the time we are card-carrying mathematicians, we also carry a label: she is an analytic number theorist, he is a computational algebraic geometer, etc., etc. Except for the very greatest of mathematicians, the possibility of stepping back, even for a glimpse of the full canvas, is hauntingly diminishing. The commonly acknowledged "last universalist", the great Henri Poincaré, died in 1912.

Of course, one should not be disheartened. Cross-links between different branches still happen from time to time; whenever a new thread is woven into the tapestry, especially when it provides an unexpected skein of thought, deep and novel mathematics is generated. These "correspondences" occur infrequently because they require either fortuitous accidents or insights far ahead of contemporaries. They need visionaries and mystics. Vladimir Arnold, with a breadth of knowledge typical of the Russian school, is one visionary who has been a proponent of "Poly-mathematics" [2], which seeks connections and correspondences.

Poly-Mathematics: A prime example of an unexpected thread, from which this book draws its inspiration, is the **McKay Correspondence** [3], discovered by the fourth author in 1980. It relates the classification of the discrete rotational symmetries in $\mathbb{R}^3$, something known to the ancient Greeks, to the classification of continuous symmetries in arbitrary dimensions, the so-called Lie groups that have been known since the last years of the nineteenth century to the early part of the twentieth. That the discovery of this correspondence had to wait so long is precisely because it spans two seemingly unrelated parts of the mathematical canvas: a colourless skein has to traverse the land of representation theory of Lie algebras, to that of the combinatorics of finite groups, a skein provided by finite graph theory. A common ground seemed to be an **A-D-E** pattern in the classification of both.

This pattern will turn out, over the decades since, to be *ubiquitous* across disciplines, many of which are still rather mysterious. It, like modularity in the Langlands Programme, or q-series in Generalised Moonshine, is a repeating motif on the tapestry of mathematics. Speaking of **Moonshine**, another observation of the fourth author, in 1978, related the until-then utterly unrelated fields of modular forms in analytic number theory to the representation theory of finite simple groups. Though the proof of the formalised conjecture of Conway and Norton was given in 1992 by Borcherds in his Fields-Medal-winning work for the so-called Monster Sporadic group, such correspondences for other finite groups remain an active programme of research. Even more strikingly, the A-D-E pattern emerges in Moonshine, giving us an evermore mysterious connection amongst connections.

Readership and Scope: Since the A-D-E pattern touches upon so much mathematics, spanning material which could be explained to a high-school student to that which still bemuses the most sophisticated of contemporary researchers, we see a duty incumbent upon ourselves to write a textbook to introduce this fascinating subject. While there are such introductions as [4, 5], aimed at a postgraduate-level mathematical audience, something which an advanced undergraduate in mathematics can understand and appreciate is, sadly, missing from the literature.

The reason for the lack of such a book is simple: it is a daunting task to explain so much mathematics, spanning so many centuries and different branches, to a novice. However, we believe that this is not only possible but also of great benefit to the young student. Indeed, what better subject matter to entice and to initiate than one which starts with Platonic solids, and yet grows rapidly to so many connections and still so filled with mystery?! Indeed, while almost all undergraduate textbooks on mathematics fall into the routine of "Introduction to X" where X is "group theory", "commutative algebra", or "algebraic geometry", etc., how refreshing it would be to have a book which allows the neophyte to have a glimpse at a large segment of the tapestry of mathematics all at once?!

Thus is the purpose of this book, to let the young students into the world of poly-mathematics, to attempt to envision mathematics as a whole, through the looking glass of the ADE pattern. We hope they will be enthralled and will one day unravel some of the enigmas of the pattern's ubiquity. We aim to have the earlier parts of the book accessible to an advanced undergraduate student with a foundation in linear algebra. We also provide a brief exposition of key results and examples in group theory that will repeatedly come up. This book could therefore also be used as a textbook that provides a contemporary take on some standard algebra topics such that it can be taught as an algebra or mathematical physics module, or as food for thought and a source of potential topics for student projects such as dissertations or honours theses. The latter parts can be used – not exhaustively, but likewise as inspiration – by beginning graduate students, to give them ideas about possible topics, problems and conjectures, as well as a brief précis of some key ideas along with references for deeper perusal. Nobody can be an expert in all of these areas, but it is our hope to provide a collection of signposts to different interesting areas and tantalising connections between them, and to inspire the next generation of mathematicians to think in creative and collaborative ways, seeking to make new connections and unify seemingly different areas of mathematics; in other words, to think like poly-mathematicians.

Finally, a note of warning. The mathematician's slogan is "Never apologise; always explain", and we have kept to this as far as we can. But we must apologise that, since some of this material is still mysterious, we cannot explain everything. We hope that some of our readers will help to do so!

Acknowledgements

Our dear friend and co-author Professor John McKay, whose observations inspired this entire discipline in mathematics, sadly passed away during the preparations of this book [1]. To fond memories of him, we dedicate this work. May he enjoy an infinity of new ideas and connections in the Elysian Fields of Mathematics.

The three authors left to bear the heart ache and the thousand natural shocks salute Trinh Vo-McKay for tirelessly taking care of John in his last days, and Liz Hurt (née McKay) for sorting out his estates, particularly in signing off the last forms relating to this publication.

We are grateful to our long-suffering partners who, especially during the COVID years, bore the brunt of our varying temporary madness induced by writing a book amidst all the chaos: PJC to Rosemary, PPD to Adrian, YHH to Lizzie, JM to Trinh.

PPD is grateful to Jeremy Butterfield for first introducing him to Vladimir Arnold and catastrophe theory, and to David Hestenes, who, when PPD was staying with him in Phoenix, introduced him to Arnold's papers on Trinities and polymathematics.

We are indebted to Dr. Jiakang Bao for many useful comments on the final draft, and to Dr. Amy Jacobsen of Cambridge University Press for her constant help throughout the project, as well as Roger Astley, Naomi Chopra and CUP LaTeX support.

Nomenclature

$\mathfrak{A}_n$ The alternating group permuting n elements, of size $n!/2$

C_n $\simeq \mathbb{Z}/n\mathbb{Z}$ is the cyclic group of size n; sometimes this is also denoted $\mathbb{Z}_n$, though we adhere to $\mathbb{Z}/n\mathbb{Z}$

$\Gamma = SL(2;\mathbb{Z})$ the modular group of linear fractional transformations of the integers

$\mathcal{A}(\mathcal{G})$ Adjacency matrix of a finite graph $\mathcal{G}$

$\mathfrak{S}_n$ The symmetric group permuting n elements, of size $n!$

$\oplus$ Direct sum between vector spaces

$\otimes$ Tensor product between vector spaces

τ and σ The golden ratio $\tau = \frac{1}{2}(1+\sqrt{5})$ and its Galois conjugate $\sigma = \frac{1}{2}(1-\sqrt{5})$

$\widehat{}$ or $\tilde{}$ Affine version of a root system/Coxeter group/Lie algebra/Lie group

$CP(m)$ The cocktail party graph on m pairs of nodes

$e_{i=1,\dots,n}$ Standard orthonormal basis of $\mathbb{R}^n$

G $= \langle R_1, R_2, \dots, R_m : f_1(R_i) = f_2(R_i) = \dots = f_M(R_i) = \mathbb{I} \rangle$ is a group generated by m elements $R_{i=1,\dots,m}$ satisfying M relations $f_{j=1,\dots,M}$

K_n The complete graph on n nodes

$K_{m,n}$ The complete bipartite graph on two sets of m and n nodes respectively

$L(\mathcal{G})$ The line graph associated to a graph $\mathcal{G}$; similarly $L(\mathcal{G}; a_1, \dots, a_n)$ is the generalised line graph

$2I$ Rotational icosahedral group of size 120, discrete subgroup of Spin(3)

$2O$ Rotational octahedral group of size 48, discrete subgroup of Spin(3)

$2T$ Rotational tetrahedral group of size 24, discrete subgroup of Spin(3)

$\mathbb{C}$ Complex numbers

$\mathbb{H}$ Quaternions

$\mathbb{R}$ Real numbers

A_3 Coxeter group/root system, T_h, full tetrahedral group of size 24, discrete subgroup of O(3)

A_n Infinite family of Coxeter groups/root systems/Lie groups

B_3 Coxeter group/root system, O_h, full octahedral group of size 48, discrete subgroup of $O(3)$

d_i Degree – invariant: degree of invariant polynomials of Coxeter groups

D_n Infinite family of Coxeter groups/root systems/Lie groups

E_6, E_7, E_8 Exceptional Coxeter groups/root systems/Lie groups of E-type

h Coxeter number

H_3 Coxeter group/root system, I_h, full icosahedral group of size 120, discrete subgroup of $O(3)$

I Rotational icosahedral group of size 60, A_5, discrete subgroup of $SO(3)$

$j(\tau)$ Klein's j-invariant for an elliptic curve

m_i Exponent – invariant: exponent in the complex eigenvalue of the Coxeter element w

O Rotational octahedral group of size 24, S_4, discrete subgroup of $SO(3)$

R Rotor/spinor: a (normalised) element of the even subalgebra of a Clifford algebra

T Rotational tetrahedral group of size 12, A_4, discrete subgroup of $SO(3)$

w Coxeter element

Dih_n is the dihedral group of size $2n$, which is the semi-direct product $C_n : C_2$, the cyclic symmetry of a regular n-gon, together with the 2-fold symmetry of flipping the n-gon over.

LHS   Left-hand side

RHS   Right-hand side

1

An Invitation

NASA's Pioneer Missions in the mid-1970s sent out probes *Pioneer 10 & 11*, which have now left our solar system carrying a famous plaque. It depicts the images of a man and woman to identify our species, the spin-flip transition of hydrogen to set a length measurement, the position of the Sun to the centre of the Galaxy and to 14 pulsars, as well as a diagram of the Solar System. These images are to serve as a calling card of humanity to possible extra-terrestrial civilisations. Professor Francis Buekenhout of Brussels suggested that the set of diagrams shown in Figure 1.1 should be included, his view being that even if an alien civilisation had very different mathematics to ours, chances are that they would have come up with at least some of the areas in which such a figure, called **ADE Diagrams**, should occur.

That these diagrams, or what Professor Terry Gannon [6] calls a "meta-pattern", consisting of 2 infinite series and 3 isolated – "exceptional" or "sporadic" – cases, should emerge in various branches ranging from algebraic geometry to representation theory to mathematical physics, places them in a central and rather mysterious position in the very structure of mathematics. As a tribute to Hilbert's famous 23 problems at the turn of the twentieth century, Professor Vladimir Arnold posed the understanding of the ubiquity of the ADE diagrams as a key problem to modern mathematics [7]. Indeed, in his lecture honouring the founding of the Clay Institute, whose Millennial Problems also echo those of Hilbert, Professor Nigel Hitchin [8] chose the subject of the three E-type diagrams.

Over the years, an increasing number of classification problems and correspondences, at first curious, then uncovered to be profound, have resulted in an ADE-type of meta-pattern:

$$\boxed{\begin{array}{l} \text{2 infinite families} \\ \text{3 exceptional cases} \end{array}} \qquad (1.1)$$

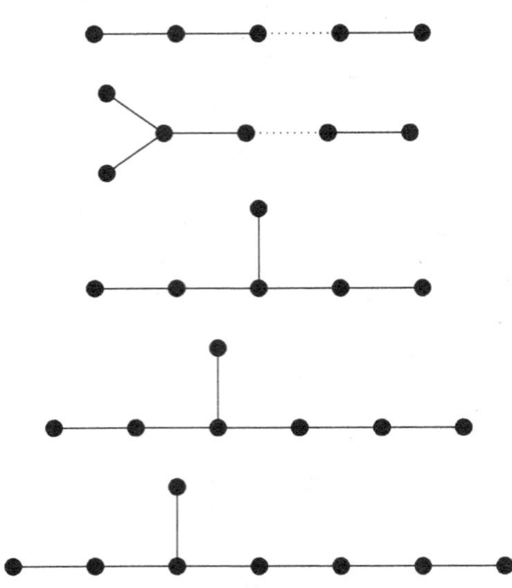

Figure 1.1 ADE diagrams – a universal pattern?

The 2 infinite families are called type A and D while the 3 exceptionals are called type E. The diagrams above, the ones which ought to be on humanity's calling card, are clearly of this form. The elevation of the diagrams to almost religious heights has suggested the usage of the word **"ADE-ology"**.

The purpose of this book is to introduce the reader to this vast and fascinating topic of ADE-ology. Whilst there have been nice reviews in the past, especially after the advent of the McKay Correspondence by the fourth author in 1979 (cf. e.g., [3, 6, 9]), an up-to-date and pedagogical treatment, suited for an advanced undergraduate or an early post-graduate, is somewhat lacking. It is our hope that the present writing can help initiate those with an appetite for webs of inter-connections into this fascinating field.

1.1 Origins

We could trace ADE-ology to as far back as the ancient Greeks' work on Platonic solids, or perhaps even earlier, if the Scottish solids are taken as indicative of awareness of the Platonic solids [10]. What are the "perfect" shapes in ordinary Euclidean space? They are the regular polygons in 2D and the five Platonic solids, each of which is comprised of specific identical regular polygons. The regular polygons in 2D are also related to the n-prisms in 3D (by

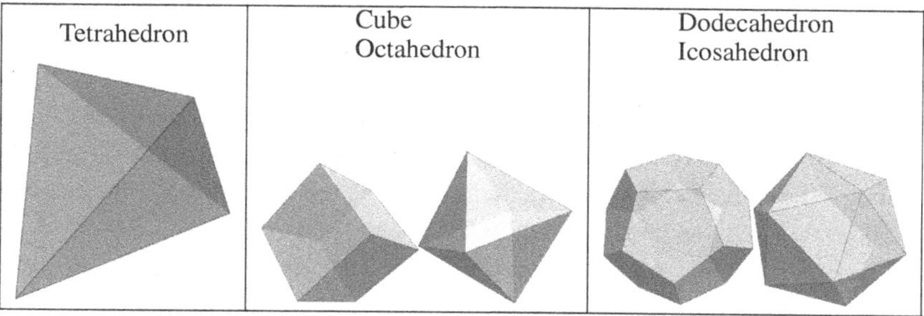

| Tetrahedron | Cube
Octahedron | Dodecahedron
Icosahedron |

Figure 1.2 The five Platonic solids.

having a trivial 3rd dimension, either via a reflection or a 2-fold rotation –
which are the same as abstract groups), which will be surprisingly interesting
later. In summary, these regular shapes are the regular n-gons and n-prisms as
well as the regular polyhedra in Figure 1.2.

We have grouped the Cube-Octahedron and the Dodecahedron-Icosahedron
together because they are graph-dual to each other in the normal sense: exchange each vertex with a face and each edge with a perpendicular edge. In
detail, if we put a vertex in the centre of each face of a regular polyhedron, and
join the new vertices in neighbouring faces by edges crossing those separating
the faces, we obtain the dual polyhedron. In particular, a polyhedron and its
dual have the same symmetry group.

Theaetetus is thought to have classified the regular solids around 400 BC.
They are however named after his contemporary Plato, who described a "cosmology" based on these five solids in his *Timaeus*: four of them correspond
to the four elements (icosahedron water, cube earth, octahedron air, tetrahedron fire) whilst the remaining dodecahedron is taken to represent the ordering principle of the universe. Euclid's *Elements* culminate in the classification
of the Platonic solids. Aristotle postulated the existence of a "fifth element"
(quintessence) as the "ether" of which the heavens are made, but unlike Plato,
he did not associate this with the dodecahedron.

Kepler also used the Platonic solids as an ordering principle of the universe
when describing the motions of the planets, with nested polyhedra predicting
the relative radii of circular orbits, before discovering his more famous laws of
elliptical planetary motion. In contemporary cosmology, there is a possibility
that despite the observed flatness of the universe, it could still be a very large
3-sphere S^3 (see Section 3.4). In fact, this 3-sphere could be tiled by 120 identical tiles, and periodic identification across the edges of these tiles yields the

Poincaré dodecahedral space model/homology sphere (i.e., the sphere modded out by the binary icosahedral group, see Section 3.4.2) [11]. There was some recent interest that the universe might actually be such a Poincaré dodecahedral space, with the non-trivial global topology potentially leaving imprints in the cosmic microwave background [12]. Whilst perhaps ultimately not being the shape of the universe, it was nonetheless interesting to reconnect the universe and Plato's old dodecahedron idea in the twenty-first century.

Since the Platonic solids are so ubiquitous, it is therefore natural to let us begin, as an invitation, with something which the reader might recall from school:

THEOREM 1.1 *There are 5 Platonic solids.*

Proof First, we recall the definition that a Platonic solid is "perfect" in the following sense: every face is an identical regular p-gon and every vertex has the same number of edges, say q, meeting. Clearly, p, q are integers at least 3.

Consider, therefore, any vertex where 2 edges from each face meet. The angle between these 2 edges is $\frac{\pi(p-2)}{p}$, being the internal angle of a regular p-gon. Since q such edges meet, we must have

$$q\frac{\pi(p-2)}{p} < 2\pi , \tag{1.2}$$

since otherwise the edges would all be in a plane (when the RHS is 2π) or bend over (when then RHS $> 2\pi$). In other words, the inequality is strict in order to have a convex body.

The above inequality rearranges to the Diophantine inequality in Egyptian fractions:

$$\frac{1}{p} + \frac{1}{q} > \frac{1}{2} , \qquad p, q \in \mathbb{Z}_{\geq 3} . \tag{1.3}$$

Now, if either $p, q \geq 6$, the other would be forced to be < 3. Thus we only need to check p, q in the range $\{3, 4, 5\}$ and find exactly 5 solutions:

(p, q)	Platonic solid
$(3, 3)$	Tetrahedron
$(3, 4)$	Octahedron
$(3, 5)$	Icosahedron
$(4, 3)$	Cube
$(5, 3)$	Dodecahedron

These are the 5 in the diagram in (1.2). In particular, the aforementioned graph duality corresponds to exchanging $(p \leftrightarrow q)$. There are more sophisticated ways to show this, using, for instance, the Euler number and Schläfli symbols for polyhedra, but the above, purely elementary argument is perhaps the most elegant. □

What we have seen above, the classification of the perfect shapes in 3D, is our first encounter with ADE-ology. We will return to this again in Section 3.1.

1.2 Dramatis Personæ

Grouped in this judicious way as in Figure 1.2, we see 3 exceptional cases. Of course, in addition to these there are 2 infinite families of objects: (i) the regular n-gons and (ii) the n-prisms. This book is about a pattern which permeates many different classification problems in seemingly utterly different branches of mathematics: the pattern we have seen in Equation (1.1). This meta-pattern is so universal that the reader is encouraged to bear in mind that firstly it might be found in any field of study, and secondly, whenever one encounters it, one should look for a potential underlying ADE structure.

Let us be more precise about the central character of our book. There will in fact be a pair of protagonists, each a graph with integer labels on the nodes, to which we shall refer, respectively as **ADE** and **Extended ADE** (also called "Affine ADE") diagrams. These are summarised in Table 1.1, with the standard notation that the affine case has a hat or tilde on top. We have also labelled the nodes with integers called Coxeter labels. What all these words mean will be be explained in the text, in particular in Sections 3.5 and 4.3.

1.3 Navigating This Book

After this invitation, we now present how this book can be used by different readerships. We have tried to keep the key material at as far as possible an elementary level. Readers who have a good grounding in linear algebra and group theory can skip Chapter 2 and directly proceed to the chapter on ADE sets and classifications (Chapter 3). This is then followed by the chapter on links between these sets known as ADE correspondences (Chapter 4). Chapter 2 however can also be used either as a refresher of these foundational topics with a plethora of relevant examples; or it can in fact also serve as an undergraduate course in these standard topics, but set in a modern context of cutting-edge topics of interest. Chapter 5 then goes far beyond elementary material to explore a loosely connected collection of much more sophisticated advanced topics of current interest, which the reader is of course at liberty to peruse or skip as

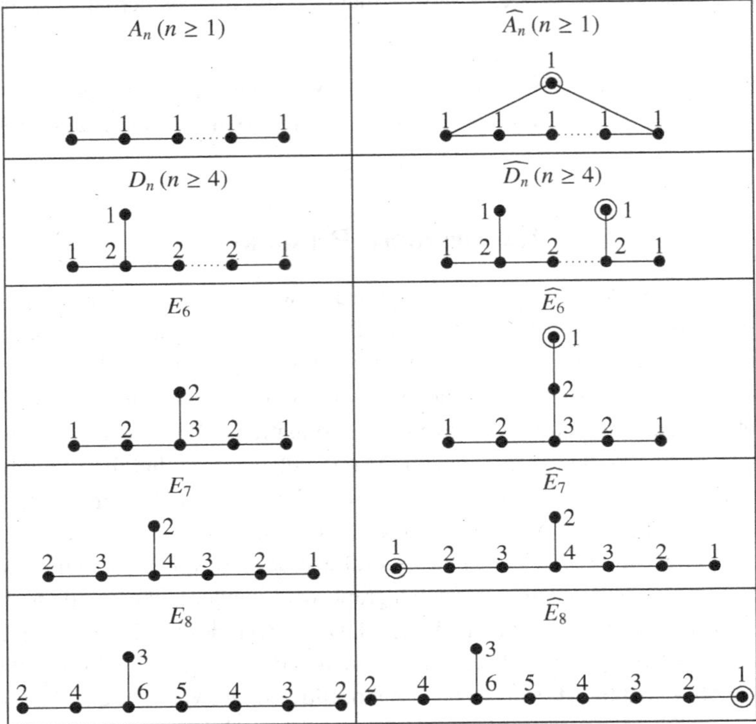

Table 1.1   *The ADE and extended (affine) ADE diagrams.*
The integer labels are the so-called Coxeter labels. The two sets of diagrams only differ in the addition of one node on the right. These affine nodes are circled explicitly. We will explain what these words mean later.

appropriate and referred to the literature for exhaustive treatments. These topics are opening up interesting areas and current research directions across a broad range of mathematics, which can act as a source of interesting dissertation projects for advanced students or inspire research projects for interested readers.

2
Algebraic Preliminaries

We don't attempt an exhaustive introduction to algebra here, though this section can be used as a foundation for a standard undergraduate course motivated by a contemporary perspective. We will introduce (or recall) central concepts and results, and give high-level conceptual connections and examples that will make an appearance again later. For those who already have some knowledge of introductory algebra, it should be a leisurely read recalling basic facts and concepts from algebra and group theory, so that we can recognise some of the characters that we will see repeatedly later and can start relating them to each other. Our introduction will be informal rather than pedantic, and we hope its vivacious pace would serve to entertain the beginning student.

2.1 Vector Spaces

A vector space is defined over a field $\mathbb{F}$ (often, but not always, the real numbers $\mathbb{R}$ or the complex numbers $\mathbb{C}$), whose elements are called *scalars*; elements of the vector space V are called *vectors*. Two vectors can be added, and a vector can be multiplied by a scalar.

Vector spaces are defined by their linear structure, and are also often called "linear spaces" for that reason. The linear maps on this space are closely linked to the set of matrices known as $GL(n; \mathbb{R})$, because linear transformations can be – and are usually – expressed as matrices acting on vectors (though we will see an alternative in Section 3.4.4). We will see $GL(n; \mathbb{R})$ again below, because these matrices actually have the structure of a group, so we will be seeing them again in Section 2.3.

Often vector spaces have structures in addition to just linearity:

Scalar product: One example is a scalar product (also called a dot or inner product), denoted $v \cdot w$, which defines a (bilinear) product of two vectors v, w in the vector space, which yields a scalar.

Algebra: Another example is a (bilinear) product of vectors that yields another vector in the same vector space, which turns this vector space into an "algebra". We will see some familiar examples such as $\mathbb{R}$, $\mathbb{C}$ and $\mathbb{H}$ (as the "normed division algebras"). Another perhaps familiar example is the "exterior algebra"; and using a scalar product on a vector space allows one to define the corresponding "Clifford algebra".

Multilinear maps are maps that are linear in several arguments and are also referred to as "tensors". A tensor can be specified in terms of components which are the values of this multilinear map acting on complete sets of basis vectors. This is often convenient for concrete calculations such as in theoretical physics (especially in general relativity). A more abstract "frame-free" way of thinking about these multilinear maps can however also be useful for shedding light on the intrinsic geometry. This brings us to two operations which we will use throughout the book:

Direct sum $\oplus$: Given vectors v in a vector space V and w in a vector space W, the direct sum $V \oplus W$ is a vector space made of ordered pairs (v, w). A familiar example is the Cartesian plane $\mathbb{R}^2 = \mathbb{R} \oplus \mathbb{R}$. The dimension of $V \oplus W$ is the sum of those of V and W.

Tensor product $\otimes$: Let v_i (respectively w_j) be a basis of the vector space V (respectively W). The basis of $V \otimes W$ is composed of all possible formal pairs (v_i, w_j). Thus the dimension of $V \otimes W$ is the product of those of V and W.

One can build larger vector spaces out of smaller vector spaces by considering the direct sum $\oplus$ or the tensor product $\otimes$.

EXAMPLE 2.1 (2D Euclidean space)   Consider $\mathbb{R}^2$ with vectors of the form $z = \begin{pmatrix} a \\ b \end{pmatrix}$. Addition of vectors is just given componentwise by

$$z_1 + z_2 = \begin{pmatrix} a_1 \\ b_1 \end{pmatrix} + \begin{pmatrix} a_2 \\ b_2 \end{pmatrix} = \begin{pmatrix} a_1 + a_2 \\ b_1 + b_2 \end{pmatrix}.$$

Let us define also a scalar multiplication of such vectors by

$$z_1 \cdot z_2 = \begin{pmatrix} a_1 \\ b_1 \end{pmatrix} \cdot \begin{pmatrix} a_2 \\ b_2 \end{pmatrix} = a_1 a_2 + b_1 b_2.$$

This product yields a scalar and gives the usual Euclidean notions of distance, lengths, angles, etc. The distance from z_1 to z_2 is $\|z_1 - z_2\|$, where $\|z\| = (z \cdot z)^{1/2}$, and the cosine of the angle between vectors z_1 and z_2 is $(z_1 \cdot z_2)/\|z_1\|.\|z_2\|$.

The symmetric or antisymmetric part of such tensor products is often of particular interest. Given the tensor power $V^{\otimes k} = V \otimes V \otimes \cdots \otimes V$ of a vector space V of dimension n, the two subspaces of most interest are

$\mathbf{Sym}^k V$: The symmetric product is the subspace of $V^{\otimes k}$ whose basis is the subset of the n^k ordered k-tuples of the basis of V, but considering any permutation to be equivalent. A simple example which illustrates this is to take V to be the vector space whose basis is composed of formal variables x and y. Then, the basis of $\mathrm{Sym}^3 V$ consists of all degree 3 monomials in these two variables, namely, $x^3, x^2 y, xy^2, y^3$. We see that the identification under permutation symmetry is reflected by the fact that ordinary multiplication of the variables is commutative (unordered). Thus, $\dim\left(\mathrm{Sym}^k V\right) = \binom{n+k-1}{k}$. (The proof of this is a non-trivial combinatorial exercise.)

$\bigwedge^k V$: On the other extreme, the antisymmetric (or exterior, or wedge) power of V has basis consisting of those of ordered k-tuples which have no repeated indices in a completely antisymmetrised way. This means that $v \wedge w = -w \wedge v$. For example, if V is dimension 3, with basis $\{v_1, v_2, v_3\}$, then $\bigwedge^3 V$ has basis $v_1 \wedge v_2 \wedge v_3$ (which, due to total antisymmetry, is equal to $\mathrm{sign}(\sigma) v_{\sigma(1)} \wedge v_{\sigma(2)} \wedge v_{\sigma(3)}$ for any permutation σ). In general, $\dim\left(\bigwedge^k V\right) = \binom{n}{k}$.

EXAMPLE 2.2 ($\mathbb{C}$ as an algebra)   Again, consider $\mathbb{R}^2$ with vectors of the form $z = \binom{a}{b}$. Addition of vectors is just as usual in $\mathbb{R}^2$, but let us now also define a multiplication of vectors by

$$z_1 \times z_2 = \binom{a_1}{b_1} \times \binom{a_2}{b_2} = \binom{a_1 a_2 - b_1 b_2}{a_1 b_2 + b_1 a_2}.$$

Then this product is again in $\mathbb{R}^2$ and thus furnishes an algebra. In fact, this is equivalent to the usual notion of complex numbers $\mathbb{C}$.

2.2 Polytopes

We used the Platonic solids as an initiation to ADE-ology. It is expedient to discuss them in more detail and formality here. We all have familiarity with a convex polygon, consisting of vertices and edges. The "convex" condition refers to the fact that each edge, if extended infinitely in both directions, must have the entire polygon "to its one side". For example, a rectangle is convex, but a 5-pointed star is not. To put this more formally, a convex polytope P has two equivalent definitions:

DEFINITION 2.3 (The Vertex Representation of a polytope)   A polytope is the convex hull of a set S of m points (vertices) $p_i \in \mathbb{R}^n$. In other words,

$$P := \mathrm{Conv}(S) = \left\{ \sum_{i=1}^{m} \alpha_i p_i \ : \ \alpha_i \geq 0, \ \sum_{i=1}^{m} \alpha_i = 1 \right\}.$$

DEFINITION 2.4 (The Half-hyperplane Representation of a polytope)   A polytope is the intersection of linear inequalities (hyperplanes) $H\underline{x} \geq \underline{b}$, where $\underline{b}$ and $\underline{x}$ are real n-vectors and H is some $m \times n$ matrix.

Since we will exclusively deal with convex polytopes, we will drop the adjective "convex" liberally.

The term **"polytope"** was coined by Alicia Boole Stott [13] as a generalisation of the words polygon (Greek for "many-sided") and polyhedron (Greek for "many-faced", or rather, "many-seated").

DEFINITION 2.5   The extremal points of P are called vertices, extremal lines, edges, and then 2-faces, 3-faces, etc., and the $(n-1)$-faces of codimension 1 are called facets. In $\mathbb{R}^2$, P is called a **polygon** and in $\mathbb{R}^3$ it is called a **polyhedron**.

Let us see Definitions 2.3 and 2.4 in action for a concrete example.

EXAMPLE 2.6 (Polygon example)   Consider the triangle given by the vertices $(0, 1)$, $(1, 0)$, and $(-1, -1)$ (see Figure 2.1). This is the vertex representation. An equivalent representation in terms of half-planes is given by $-x - y \geq -1, 2x - y \geq -1$, and $-x + 2y \geq -1$. Note that one can also rewrite this in terms of the scalar product, so that the edge equations can be written as

$$(x, y) \cdot (-1, -1) = -1,$$

$$(x, y) \cdot (2, -1) = -1,$$

$$(x, y) \cdot (-1, 2) = -1.$$

The rotational symmetries of different polytopes P comprise **finite groups** G, which act as rotations in $\mathbb{R}^n$, i.e., (special) orthogonal transformations on the n coordinates of $\mathbb{R}^n$: in other words, $n \times n$ real matrices M such that $M^T M = \mathbb{I}_n$. The discrete nature of the polytope P necessarily means that G is finite. It is therefore instructive to remind the reader of a little group theory.

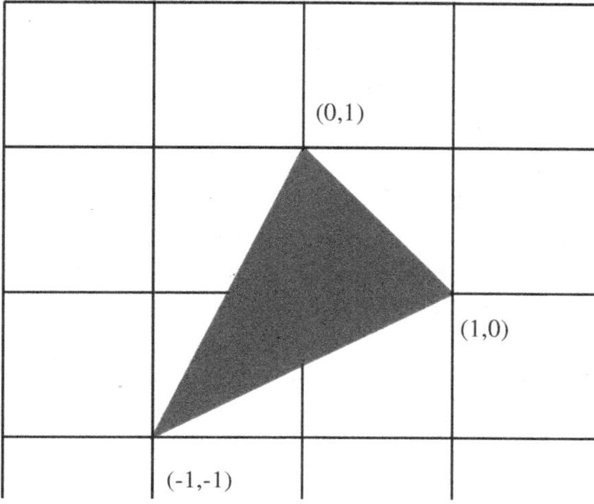

Figure 2.1 A simple triangle given in terms of its three vertices, as well as via three edge equations describing its bounding half-planes.

2.3 Groups

We have seen equilateral triangles as building blocks of the tetrahedron, octahedron and icosahedron, as well as their ability to tile the plane (Figure 2.2). This is because as a regular 3-gon, it has a certain set of symmetries to begin with, which allow (and are part of) these larger symmetry structures.

Figure 2.2 An equilateral triangle given in terms of its three vertices $(0, 0)$, $(1, 0)$ and $\frac{1}{2}(1, -\sqrt{3})$ has particular symmetries: rotations around its centre as well as reflections in the diagonals. Combinations of such symmetries are again symmetries of the equilateral triangle.

The equilateral triangle (see Figure 2.2) doesn't change ("is invariant") under rotations by 0, 120 or 240 degrees around its centre. The corresponding symmetry operations can be denoted, e.g., e, R_1, R_2. It also stays the same under reflections in its three diagonals (denoted by S_1, S_2, S_3). Combining these symmetry operations will yield another symmetry operation; in fact, it will yield one that we have already identified amongst the 6. The property that one can combine these symmetry operations and they stay within the set of all possible symmetry operations ("the set is closed" under combination) is one of the main properties of the more formal definition of a group that is meant to capture this structure: A group $G = (S, \circ)$ is a set S (e.g., of symmetry operations) with a binary combination (or "multiplication") operation $\circ$ such that closure as well as the following other properties are satisfied:

DEFINITION 2.7 (Group)   A *group* $(G, \circ)$ is a set G with a binary operation $\circ$ such that the following set of axioms is satisfied:

  (i) *Closure:* If $a, b \in G \Rightarrow a \circ b \in G$, $\forall a, b \in G$
  (ii) *Associativity:* $(a \circ b) \circ c = a \circ (b \circ c)$, $\forall a, b, c \in G$
  (iii) *Existence of an identity element e:* $a \circ e = e \circ a = a$
  (iv) *Existence of inverse elements:* $\forall a \in G$, $\exists a^{-1} \in G : a \circ a^{-1} = a^{-1} \circ a = e$.

If G is a finite set, then the number of elements in G (i.e., the cardinality of the set) is called the "order" of the group G (or simply the "size"). For a group element g the smallest integer such that $g^n = e$ is called the "order" of the group element g.

Sometimes one has concrete objects such as a set of matrices with a concrete multiplication like matrix multiplication to do group computations with. A slightly more abstract approach is to only have certain group elements, called *generators* that can generate the whole group via multiplication, subject to a set of *relations*. Multiplying such generators together creates *words* in these generators. Without any further information, this generates an infinite group, called the *free group* in these generators. However, relations reduce this number of possible words, often to something finite, by telling us which words are synonymous to each other, or to the identity.

EXAMPLE 2.8 (Quaternion group and icosahedral group)   Let us assume that we have the generators $1, i, j, k$. In principle, we could form infinitely many words out of these. However, let us say that i, j, k are imaginary units (i.e., $i^2 = j^2 = k^2 = -1$) and that furthermore $ij = k$. This actually reduces the number of possible words to only 8 words! This group of 8 elements is called the quaternion group. We will see more concrete examples of this below.

Similarly, the free group in the three generators R, S, T is infinite, but with the below relations it reduces to a group of order 60

$$\langle R, S, T \mid RST = R^2 = S^3 = T^5 = e \rangle.$$

This group is actually the group of rotational symmetries of the icosahedron, and is called the icosahedral group.

The above notation of listing the generators and the relations amongst them is called the **presentation** of a group, to whose discussion we shall soon turn.

2.3.1 Discrete Groups

Let us follow these two examples and begin with finite groups to warm up. A simple way of thinking about these is to have a finite collection of matrices that one can show to satisfy the group axioms. Matrix groups are very simple to work with. We will get a bit of practice with them below, and see the link with representation theory. But we will also see other examples of groups in terms of generators and relations, or as Clifford multivectors under Clifford multiplication. We start with the simplest non-trivial example.

EXAMPLE 2.9 Take G as the set $\{1, -1\}$ under usual multiplication. This gives the group multiplication table

	1	−1
1	1	−1
−1	−1	1

One could also consider the group H consisting of $\{0, 1\}$ with addition modulo 2. This gives the group multiplication

	0	1
0	0	1
1	1	0

The identity elements are multiplicative 1 in G and additive 0 in H. One can see that these groups are "essentially the same". If we had a mapping $\Phi : G \Rightarrow H$ that set $\Phi(1) = 0$ and $\Phi(-1) = 1$, this maps the group multiplication tables into each other. We call the groups G and H isomorphic, and consider them just different incarnations of the same abstract group $\mathbb{Z}_2$.

Even this simple example prompted thought about when groups are "essentially the same", and mappings between groups, i.e., mappings between the sets that "respect the group multiplication law". This motivates the following definitions:

DEFINITION 2.10 $(G_1, \circ_1)$ and $(G_2, \circ_2)$ are called *homomorphic* if there is a map Φ from $(G_1, \circ_1)$ to $(G_2, \circ_2)$ that respects the group multiplication law, i.e., such that

$$\Phi(x \circ_1 y) = \Phi(x) \circ_2 \Phi(y); \quad \text{for all } x, y \in G_1. \tag{2.1}$$

An antihomomorphism is analogous except that it reverses the order of the multiplication on the RHS.

The particular case of groups being "essentially the same" is encapsulated in the definition of an isomorphism as a special case when this homomorphic mapping is bijective:

DEFINITION 2.11 $(G_1, \circ_1)$ and $(G_2, \circ_2)$ are called *isomorphic* if there is a bijection (one-to-one map) Φ from $(G_1, \circ_1)$ to $(G_2, \circ_2)$ such that

$$\Phi(x \circ_1 y) = \Phi(x) \circ_2 \Phi(y); \quad x, y \in G_1 \tag{2.2}$$

holds. Φ is then called a *group isomorphism*, and we write $G_1 \cong G_2$.

Another small group that we saw above in Example 2.8 in terms of generators and relations (and shall encounter again soon) is that of the quaternions. We will meet them in their guise as a normed division algebra in Section 3.1.3. They also have a profound geometric interpretation as being related to rotations in three dimensions (Section 3.4). But for now we are just interested in their incarnation as a group.

EXAMPLE 2.12 (Quaternion group)   Consider the group G generated by

$$i = \begin{pmatrix} 0 & 1 & 0 & 0 \\ -1 & 0 & 0 & 0 \\ 0 & 0 & 0 & 1 \\ 0 & 0 & -1 & 0 \end{pmatrix} \text{ and } j = \begin{pmatrix} 0 & 0 & 1 & 0 \\ 0 & 0 & 0 & -1 \\ -1 & 0 & 0 & 0 \\ 0 & 1 & 0 & 0 \end{pmatrix}.$$

This means that one uses group multiplication (here matrix multiplication) to generate more and more elements, until the set closes.

Concretely: define $k = ij$; then there are 8 different matrices (group elements) that can be generated by matrix multiplication, namely $\pm 1, \pm i, \pm j, \pm k$, as we asserted above. Their multiplication behaviour is summarised in the multiplication table, Table 2.1. The 8 matrices define a group of order 8, called the quaternion group Q.

Q	1	-1	i	$-i$	j	$-j$	k	$-k$
1	1	-1	i	$-i$	j	$-j$	k	$-k$
-1	-1	1	$-i$	i	$-j$	j	$-k$	k
i	i	$-i$	-1	1	k	$-k$	$-j$	j
$-i$	$-i$	i	1	-1	$-k$	k	j	$-j$
j	j	$-j$	$-k$	k	-1	1	i	$-i$
$-j$	$-j$	j	k	$-k$	1	-1	$-i$	i
k	k	$-k$	j	$-j$	$-i$	i	-1	1
$-k$	$-k$	k	$-j$	j	i	$-i$	1	-1

Table 2.1   *Multiplication table for the quaternion group Q.*

Our first example of a polyhedral group will be the rotational symmetry group of a tetrahedron. This is a symmetry group of order 12. We will see it again later in Section 3.1; we will see its close relative, the full tetrahedral group (including reflections) of order 24 in Section 3.2; and we will meet its binary version, the binary tetrahedral group, also of order 24, in the context of spinors and rotations in Section 3.4, as well as in the McKay correspondence in Section 4.5.

EXAMPLE 2.13 (Tetrahedral group)   The following set of 12 matrices forms a group under multiplication, the rotational tetrahedral group. One can similarly produce a multiplication table, but for larger groups these get quite unwieldy.

$$\begin{pmatrix} 1 & 0 & 0 \\ 0 & 1 & 0 \\ 0 & 0 & 1 \end{pmatrix}, \begin{pmatrix} 0 & 0 & 1 \\ 1 & 0 & 0 \\ 0 & 1 & 0 \end{pmatrix}, \begin{pmatrix} 0 & 0 & -1 \\ 1 & 0 & 0 \\ 0 & -1 & 0 \end{pmatrix}, \begin{pmatrix} 0 & 1 & 0 \\ 0 & 0 & 1 \\ 1 & 0 & 0 \end{pmatrix},$$

$$\begin{pmatrix} 1 & 0 & 0 \\ 0 & -1 & 0 \\ 0 & 0 & -1 \end{pmatrix}, \begin{pmatrix} 0 & 1 & 0 \\ 0 & 0 & -1 \\ -1 & 0 & 0 \end{pmatrix}, \begin{pmatrix} 0 & -1 & 0 \\ 0 & 0 & -1 \\ 1 & 0 & 0 \end{pmatrix}, \begin{pmatrix} -1 & 0 & 0 \\ 0 & 1 & 0 \\ 0 & 0 & -1 \end{pmatrix},$$

$$\begin{pmatrix} 0 & -1 & 0 \\ 0 & 0 & 1 \\ -1 & 0 & 0 \end{pmatrix}, \begin{pmatrix} 0 & 0 & -1 \\ -1 & 0 & 0 \\ 0 & 1 & 0 \end{pmatrix}, \begin{pmatrix} 0 & 0 & 1 \\ -1 & 0 & 0 \\ 0 & -1 & 0 \end{pmatrix}, \begin{pmatrix} -1 & 0 & 0 \\ 0 & -1 & 0 \\ 0 & 0 & 1 \end{pmatrix}.$$

In fact, the quaternion group (in its guise as "something to do with rotations") is our first example of a binary double cover.

EXAMPLE 2.14 (Binary double cover)   Consider the group H consisting of

$$e = \begin{pmatrix} 1 & 0 & 0 \\ 0 & 1 & 0 \\ 0 & 0 & 1 \end{pmatrix}, \tilde{\imath} = \begin{pmatrix} -1 & 0 & 0 \\ 0 & -1 & 0 \\ 0 & 0 & 1 \end{pmatrix}, \tilde{\jmath} = \begin{pmatrix} -1 & 0 & 0 \\ 0 & 1 & 0 \\ 0 & 0 & -1 \end{pmatrix}, \tilde{k} = \begin{pmatrix} 1 & 0 & 0 \\ 0 & -1 & 0 \\ 0 & 0 & -1 \end{pmatrix},$$

the group of 180-degree rotations that we have seen as part of the tetrahedral group above (it is easy to check that this is indeed a group). Let us try to find a homomorphism from the quaternions to this group: Assign $\Phi\colon \pm 1 \to 1, \pm i \to \tilde{\imath}, \pm j \to \tilde{\jmath}, \pm k \to \tilde{k}$. This is a homomorphism with i, j, k, etc. "doubly covering" the 180-degree rotations. This property is associated with "spinors" – objects which perform rotations – which are closely connected with quaternions in 3D, and which we shall encounter again a lot in that context. In general, then, a double cover of a group G is a group H having a two-to-one homomorphism onto G.

The rotational polyhedral symmetries are given by **finite groups** G, which act as rotations in $\mathbb{R}^n$, i.e., orthogonal transformations on the n coordinates of $\mathbb{R}^n$. (We will encounter orthogonal transformations in a different context again shortly.) In other words, they are $n \times n$ real matrices M such that $M^T M = \mathbb{I}_n$. The discrete nature of P of course means that G is finite such that they are finite groups like the ones we have discussed above. Every finite group is finitely generated with a finite number of relations and we denote it as follows:

DEFINITION 2.15   A finite group G with identity element e can be given by a *presentation*

$$G = \langle R_1, R_2, \ldots, R_m : f_1(R_i) = f_2(R_i) = \cdots = f_M(R_i) = e \rangle$$

if G is the "largest" or free-est group generated by m elements (generators) $R_{i=1,\ldots,m}$ satisfying M relations $f_{j=1,\ldots,M}$. (More precisely, this means that if H is any group generated by m elements $S_1, \ldots, S_m$ satisfying these relations, then there is a homomorphism from G onto H carrying R_i to S_i for $i = 1, \ldots, m$.)

(We should add a cautionary note here. Presentations of groups are not always easy to work with, and indeed it is an undecidable problem whether a group given by a presentation is the trivial group with one element. If you need convincing, try the second exercise at the end of this chapter.)

Some of the most standard groups we will use in this book are the following:

- $C_n \simeq \mathbb{Z}/n\mathbb{Z}$ is the cyclic group of size n (we will sometimes use "size" rather than "order" for finite groups so as to avoid the proliferation of this overused word);

- $\mathfrak{S}_n$ is the symmetric group consisting of all permutations of the set $\{1,\ldots,n\}$ of n elements, of size $n!$;

- $\mathfrak{A}_n$ is the alternating group consisting of all *even* permutations of n elements (those for which the difference between n and the number of disjoint cycles is even), of size $n!/2$;

- Dih_n is the dihedral group of size $2n$, which is generated by $C_n : C_2$, the cyclic symmetry of a regular n-gon, together with the 2-fold symmetry of flipping the n-gon over.

When one has two groups, one can define a larger group using the smaller ones as building blocks. There are two main ways to construct such a product group:

DEFINITION 2.16 (Product groups)   Given groups G (with operation $\circ_G$) and H (with operation $\circ_H$), then the set for the product group is given by the Cartesian product, $G \times H$, i.e., the ordered pairs (g,h), for $g \in G$ and $h \in H$. The *direct product* $G \times H$ endows this set with the following group multiplication

$$(g_1, h_1) \circ (g_2, h_2) = (g_1 \circ_G g_2, h_1 \circ_H h_2).$$

If one has slightly more structure, one can also define the *semi-direct product* $G \rtimes H$ of two groups. Suppose $\phi : H \to \mathrm{Aut}(G)$ is a homomorphism sending elements $h \in H$ to automorphisms ϕ_h of G (automorphisms are self-isomorphisms, i.e., isomorphisms from an algebraic object to itself). Then the group $G \rtimes_\phi H$ is the group generated via the *twisted* multiplication law given by

$$(g_1, h_1) \circ (g_2, h_2) = (g_1 \circ_G \phi_H(g_2), h_1 \circ_H h_2).$$

Note that the direct product is essentially a trivial special case of this for an untwisted product given by a trivial homomorphism.

This allows one to construct larger and larger groups out of building blocks. Conversely, it allows a large group to be decomposed in terms of smaller building blocks, as we now describe.

The smallest building blocks here are called *simple*, which means groups that have no non-trivial normal subgroups (normal subgroups are subgroups invariant under conjugation[1]). This is because for a larger group one can sometimes use normal subgroups to decompose it as a semi-direct product (in

---

[1] A subgroup N of a finite group G is called *normal* if it is a union of some of the conjugacy classes of G; equivalently, if $n \in H$, then the conjugacy class of N is contained in H. Normal subgroups are important because of their connection with homomorphisms, and we will introduce these shortly in Section 2.6.

reverse to the above). Thus simple groups are the analogues to prime numbers when it comes to building, and classifying, groups!

The endeavour to classify finite simple groups has been a monstrous effort in the twentieth century, leading to classification results filling several large tomes. These finite simple groups also come in several infinite families as well as some 26 exceptional ("sporadic") cases. We will briefly hint at these sporadic groups later in Section 5.1, but only in the context of a potential ADE pattern in this field.

2.3.2 Continuous Groups

So far, we had finite sets of matrices that satisfied the group axioms. Having an infinite set of matrices to work with may seem daunting, but is actually very straightforward when there are continuous parameters to work with. (Indeed, the most difficult group axiom to verify, the associative law, is automatically satisfied by matrices.) The elementary example is just the circle: the unit complex numbers, or rotations in the plane around the origin. We will formalise some of the notation shortly.

As a historical note, the concept of finite groups appeared first, notably in the work of Galois in studying the permutations of roots of polynomials. Decades later, Sophus Lie started systematically studying continuous transformations, giving rise to continuous groups, now called Lie groups in his honour.

EXAMPLE 2.17 (The group U(1))   Consider the group G consisting of unit complex numbers, i.e., those forming a circle. By Euler's formula we can write these as $e^{i\phi}$. Group multiplication is therefore just $e^{i\phi_1} e^{i\phi_2} = e^{i(\phi_1 + \phi_2)}$. Geometrically they are just combinations of rotations in the plane.

We can take G as SO(2), the set of special orthogonal 2×2 matrices O acting on the plane $\mathbb{R}^2$, i.e., $\det O = 1$ and $OO^T = 1_2$. We can also think of G as U(1), the set of unitary 1×1 matrices U, i.e., matrices U with complex entries such that $UU^\dagger = 1$, where the dagger denotes complex conjugation combined with transposition.

O and U have to be of the form $O = \left(\begin{smallmatrix} \cos\phi & -\sin\phi \\ \sin\phi & \cos\phi \end{smallmatrix} \right)$ and $U = e^{i\phi}$. Both describe rotations in the plane: one in the real plane and one in the complex plane. O acts on a real 2D vector $(x, y)^T$, whilst U acts on a complex 1D vector z, but this is $z = x + iy$, so "essentially the same" preserving $x \cdot x = z\bar{z} = x^2 + y^2$. This is another example of a group isomorphism, this time for continuous groups: $SO(2) \cong U(1)$.

These groups are simply continuous groups of which the finite, discrete rotation groups that we studied above are subsets (and in fact, also subgroups). The most familiar is the set of rotational symmetries of $\mathbb{R}^3$, i.e., the set of orthogonal 3×3 real matrices. This readily generalises to $\mathbb{R}^n$:

$$O(n) := \left\{ M \in \mathrm{Mat}_n(\mathbb{R}) \ : \ M^T M = \mathbb{I}_n \right\}. \tag{2.3}$$

We will use $\mathrm{Mat}_n(\mathbb{F})$ to mean $n \times n$ matrices over a field $\mathbb{F}$, with $\mathbb{F}$ typically being $\mathbb{R}$ and $\mathbb{C}$. Orthogonal matrices automatically have determinant ± 1. We can also restrict to orientation-preserving rotations by imposing that the determinant be positive. These are also the special orthogonal groups:

$$SO(n) := \{ M \in O(n) \ : \ \det(M) = +1 \}. \tag{2.4}$$

We therefore have some typical continuous groups, the special orthogonal and special unitary groups $SO(n)$ and $SU(n)$:

$$SO(n) := \left\{ M \in \mathrm{Mat}_n(\mathbb{R}) \ : \ M^T M = \mathbb{I}_n, \ \det(M) = +1 \right\},$$

$$SU(n) := \left\{ M \in \mathrm{Mat}_n(\mathbb{C}) \ : \ \overline{M}^T M = M^\dagger M = \mathbb{I}_n, \ \det(M) = +1 \right\}.$$

We also define the general linear group

$$GL(n; \mathbb{R}) := \{ M \in \mathrm{Mat}_n(\mathbb{R} \ : \ \det(M) \neq 0 \}$$

of invertible matrices, as well as the special linear group

$$SL(n; \mathbb{R}) := \{ M \in GL(2; \mathbb{R}) \ : \ \det(M) = +1 \}.$$

These are all easily seen to be continuous groups. In a wider context, we can therefore recast the problem of finding the regular symmetries of $\mathbb{R}^3$ in Section 3.1 as the problem of classification of discrete, finite, subgroups of $O(3)$. These types of matrix groups are continuous groups. In fact, the continuity gives them a particularly nice "smooth" structure: that of a manifold. We won't go into manifolds here, but the idea is that manifolds are spaces that look locally like $\mathbb{R}^n$ but can be curved on a global scale. Think of the sphere, which looks locally like $\mathbb{R}^2$ – but the globe is topologically quite different from a flat earth!

Groups that also have such a nice manifold structure are called Lie groups. We will encounter them again in Section 3.5, and they are also closely linked to the so-called Lie algebras. A lot of the original ADE observations were historically made in this Lie context. Many ADE-type structures to do with these are however also hidden in the somewhat more elementary polyhedral groups and root systems, so we will focus on these first, and introduce Lie theory a little later. So we will stick with thinking about these objects as continuous groups for now, and defer thinking about the additional manifold structure that also makes them into Lie groups till later.

2.4 Matrices and Groups

In linear algebra, we are often interested in simultaneous systems of equations such as

$$x + 3y + 4z = 10,$$
$$x + 2y + 6z = 12,$$
$$3x + 3y + 8z = 26,$$

which can be written more succinctly in matrix form as

$$\begin{pmatrix} 1 & 3 & 4 \\ 1 & 2 & 6 \\ 3 & 3 & 8 \end{pmatrix} \times \begin{pmatrix} x \\ y \\ z \end{pmatrix} = \begin{pmatrix} 10 \\ 12 \\ 26 \end{pmatrix}. \tag{2.5}$$

Linear algebra in a way is a systematic theory to solve such equations. One often gets such sets of equations from linear maps or linear functions: we want to map a vector, a point in space, to some transformed point, e.g., via a rotation. The components of the original and transformed vector with respect to a basis give exactly such a set of simultaneous equations:

$$\begin{pmatrix} 1 & 3 & 4 \\ 1 & 2 & 6 \\ 3 & 3 & 8 \end{pmatrix} \times \begin{pmatrix} x \\ y \\ z \end{pmatrix} = \begin{pmatrix} x' \\ y' \\ z' \end{pmatrix}. \tag{2.6}$$

Therefore, matrices are essentially encoding linear transformations (i.e., the general linear group $GL(n; \mathbb{R})$ above), although they can also have other purposes. Conversely, linear transformations can be written as matrices – although there are also other possibilities (we will encounter an example later, in the context of Clifford algebras).

Diagonal matrices just act on the different components as rescalings. Diagonalisation is the next best thing: one finds a new basis (given by the eigenvector[2]), in which the linear transformation acts as a simple rescaling (given by the eigenvalues). Eigendecomposition is essentially a similarity transformation (see below) that describes the change of basis to the basis of eigenvectors, where the simple rescalings act. In many interesting application cases (e.g., symmetric or Hermitian matrices) the eigenvectors are orthogonal.

---

[2] The terms "eigenvector" and "eigenvalue", a mixture of English and German, used to be translated into English as "characteristic vector/value" or "proper vector/value". In older books one sees the term "latent root", indicating that the eigenvectors are hidden in the matrix and can be brought out by applying a certain procedure, much as the latent image on a photographic plate was revealed by developing the plate.

Eigentheory is typically covered in an introductory linear algebra course, and we will skip the details here. At this point, we just cover one theorem, the *Perron–Frobenius theorem*, which will be of interest later. Earlier work by Perron concerns positive matrices:

THEOREM 2.18 *For an $n \times n$ positive matrix A, there is an eigenvalue r whose absolute value is bigger than all others, whose corresponding eigenspace is one-dimensional, and whose eigenvector v_{PF} has all positive entries.*

This can in fact be slightly generalised, e.g., to non-negative and positive semidefinite matrices (Frobenius). (A matrix A is positive semidefinite if $v \cdot Av \geq 0$ for any vector v.)

THEOREM 2.19 *For an $n \times n$ real symmetric, positive semidefinite and indecomposable matrix with non-positive off-diagonal entries A, the eigenvalues are real and non-negative, the null space has dimension ≤ 1 and the largest eigenvalue r has a corresponding one-dimensional eigenspace and an eigenvector with all (strictly) positive entries.*

These Perron–Frobenius results will be important in various contexts later, in particular in the construction of the Coxeter plane in Section 3.2.4 and Smith's theorem, Theorem 3.9.

2.5 Equivalence Relations and Conjugacy Classes

As we have seen above with group multiplication tables, working with many group elements can become very cumbersome. It would help if we could group "similar" elements together in some way, so as to partition the group into more manageable chunks. Conjugacy classes are one way of dividing up group elements in a useful way. Often these have similar properties within each class, which are characterised by invariants (or class functions). Simple invariants, as we shall see, could be the trace, order or determinant of a matrix. That way, e.g., rotations could be separated from reflections. Another common way of partitioning a group is via cosets, although this requires a subgroup H as well as the original group G.

Such a partition would have the following useful properties. It would partition the group into non-intersecting pieces ("classes"), so that each group element falls into exactly one class. And secondly, the elements in each such "class" relate to each other in a sensible way. These ideas are formalised in the notion of an "equivalence relation".

DEFINITION 2.20 (Equivalence relation)   A binary relation $\sim$ on a set X is said to be an equivalence relation if and only if it is reflexive, symmetric and transitive. That is, for all a, b and c in X:

- *Reflexivity:* $a \sim a$
- *Symmetry:* if $a \sim b$ then $b \sim a$
- *Transitivity:* if $a \sim b$ and $b \sim c$ then $a \sim c$.

The job of an equivalence relation is to partition a set. (A partition of a set X is a collection of non-empty, pairwise disjoint subsets of X whose union is X; informally, a division of X into non-empty parts.) Let $\sim$ be an equivalence relation on X. We denote by $[a]$ the equivalence class defined by a, that is,

$$[a] = \{x \in X : x \sim a\}.$$

Now by reflexivity, $a \in [a]$; also $[a] = [b]$ if and only if $a \sim b$. Hence, if $c \in [a] \cap [b]$, then $a \sim c \sim b$, and so $a \sim b$ by transitivity, and $[a] = [b]$. This proves that distinct equivalence classes are disjoint. Since every element $a \in X$ lies in some equivalence class, namely $[a]$, we have a partition. Conversely, if we are given a partition of X, define a relation $\sim$ by putting $x \sim y$ if some part of the partition contains x and y; this is an equivalence relation whose equivalence classes form the given partition.

Let us return to our motivating example for this definition by defining a concrete equivalence relation for the groups we are interested in. Here, this equivalence relation is called "conjugation", and the equivalence classes are called "conjugacy classes":

DEFINITION 2.21 (Conjugation)   An element $g' \in G$ is called *conjugate to another element* $g \in G$ if there exists an element $h \in G$ such that

$$g' = hgh^{-1}.$$

EXAMPLE 2.22 (Conjugacy as an equivalence relation)   Let us check the three equivalence relation axioms in turn:

- *Reflexivity* ($a \sim a$): for the identity (i.e., $h = e$) we have $g' = hgh^{-1} = ege^{-1} = ge = g$. So g is indeed conjugate to itself.
- *Symmetry* (if $a \sim b$ then $b \sim a$): If g' is conjugate to g, i.e., $g' = hgh^{-1}$, then by left multiplying by h^{-1} and right multiplying by $(h^{-1})^{-1} = h$ (since $h^{-1} \in G$), g must conversely be conjugate to g': $g = (h^{-1})g'(h^{-1})^{-1} = (h^{-1})g'h$ (conjugacy with respect to h^{-1}).

- *Transitivity* (if $a \sim b$ and $b \sim c$, then $a \sim c$): if $g' = h_1 g h_1^{-1}$ and $g'' = h_2 g' h_2^{-1}$ for some $h_1, h_2 \in G$), then g is also conjugate to g'' since

$$g'' = h_2 g' h_2^{-1} = h_2 h_1 g h_1^{-1} h_2^{-1} = (h_2 h_1) g (h_2 h_1)^{-1},$$

where we are using $(h_2 h_1^{-1})^{-1} = h_1^{-1} h_2^{-1}$ for inverses (the inversion map is an anti-automorphism). Since $h_2 h_1 \in G$ by group closure, g is conjugate to g''.

So there is a partition of a group into disjoint sets of mutually conjugate elements.

DEFINITION 2.23 (Conjugacy class)   A *conjugacy class* of a group G is a set of mutually conjugate elements of G. It is an equivalence class of the equivalence relation "conjugation".

The partition into conjugacy classes can be found similar to the group multiplication table: by taking any element $g \in G$ and forming the set $\{hgh^{-1} | h \in G\}$ for all other group elements. Efficient algorithms, however, use the transitivity and partition properties of equivalence relations, so that one does not need to do all such pairings in practice.

PROPOSITION 2.24 (Conjugacy class of the identity)   *The identity e is always in a conjugacy class on its own since $g' = heh^{-1} = hh^{-1} = e$ so $g' = e$ for all h.*

We note that conjugacy classes do not need to be of the same size.

EXAMPLE 2.25 (Classes of the group $\mathfrak{S}_3$ of symmetries of an equilateral triangle)   In order to form the pairwise combinations of the six group elements $\{e, R_1, R_2, S_1, S_2, S_3\}$ (akin to the group multiplication table), we first compute the group multiplication table $g_1 \circ g_2$, and then use this to compute the combinations hgh^{-1}:

$\circ$	e	R_1	R_2	S_1	S_2	S_3
e	e	R_1	R_2	S_1	S_2	S_3
R_1	R_1	R_2	e	S_2	S_3	S_1
R_2	R_2	e	R_1	S_3	S_1	S_2
S_1	S_1	S_3	S_2	e	R_2	R_1
S_2	S_2	S_1	S_3	R_1	e	R_2
S_3	S_3	S_2	S_1	R_2	R_1	e

Now for the results of conjugation (g: rows, h: columns):

hgh^{-1}	e	R_1	R_2	S_1	S_2	S_3
e	e	e	e	e	e	e
R_1	R_1	R_1	R_1	R_2	R_2	R_2
R_2	R_2	R_2	R_2	R_1	R_1	R_1
S_1	S_1	S_3	S_2	S_1	S_3	S_2
S_2	S_2	S_1	S_3	S_3	S_2	S_1
S_3	S_3	S_2	S_1	S_2	S_1	S_3

One can summarise these results as

- $heh^{-1} = e$ for all $h \in G$, so $e \sim e \Rightarrow \{e\}$ is a conjugacy class;
- $hR_1h^{-1} = R_1$ for $h \in \{e, R_1, R_2\}$ and $hR_1h^{-1} = R_2$ for $h \in \{S_1, S_2, S_3\}$, so $R_1 \sim R_2$ and likewise $hR_2h^{-1} = R_2$ for $h \in \{e, R_1, R_2\}$ and $hR_2h^{-1} = R_1$ for $h \in \{S_1, S_2, S_3\}$, so of course also $R_2 \sim R_1$); so $\{R_1, R_2\}$ is a conjugacy class;
- $hS_1h^{-1} \in \{S_1, S_2, S_3\}$ for all $h \in G$, so $S_1 \sim S_2 \sim S_3$, and so $\{S_1, S_2, S_3\}$ is a conjugacy class.

Thus there are three conjugacy classes for the group $\mathfrak{S}_3$: $\{e\}$, $\{R_1, R_2\}$ and $\{S_1, S_2, S_3\}$.

EXAMPLE 2.26 (Quaternion group) $\{1\}$ is always in a class on its own, and $\{-1\}$ is of course also in a class on its own since it commutes with everything. i, j, k have inverses $-i, -j, -k$. Choosing j as the group element to be "sandwiched" next we have $-iji = -ki = -j$, $-jjj = j$ and $-kjk = -ki = -j$, i.e., $\{j, -j\}$ is one class. Analogously, $\{i, -i\}$ and $\{k, -k\}$ are also conjugacy classes (which can be expected because of the cyclic symmetry of i, j, k). Thus there are 5 conjugacy classes in total, partitioning the group.

EXAMPLE 2.27 (Tetrahedral group)   The tetrahedral group also has 4 conjugacy classes: the identity, two classes of rotations by 120 degrees, clockwise and counterclockwise, around the vertices and faces, and the three rotations by 180 degrees (around the edges of the tetrahedron) that we saw above in the context of the quaternion double cover. One can see how these conjugacy classes are grouped together by "geometric meaning".

Let us briefly summarise these conjugacy partition results and make a start with identifying class invariants:

THEOREM 2.28 (Group decomposition into conjugacy classes)   (i) *Every element of a group G is in a unique conjugacy class of G.*

(ii) *No element of G can be a member of two conjugacy classes of G, i.e., conjugacy classes are disjoint and the underlying set of group elements is the disjoint union of the conjugacy classes.*

(iii) *The identity element e always forms a class on its own.*

(iv) *The order of group elements is an invariant of each conjugacy class.*

Proof The first two parts follow from the correspondence between equivalence relations and partitions, and the fact that conjugacy is an equivalence relation.

(iii) For every $h \in G$, $heh^{-1} = e$ so e is in a class by itself.

(iv) We recall that the order n of an element g is the smallest integer such that $g^n = e$. For any conjugate g' of g ($g' = hgh^{-1}$), we have that

$$g'^n = \underbrace{(hgh^{-1})(hgh^{-1})\dots(hgh^{-1})}_{n \text{ times}} = hg^n h^{-1} = heh^{-1} = hh^{-1} = e,$$

so g' is also of order n and thus all members of a conjugacy class are of the same order, which is a class invariant. $\qquad\square$

THEOREM 2.29 *If G is an abelian (i.e., commutative) group, then every element of G forms a class on its own.*

Proof Any $g, h \in G$ commute, $hg = gh$. Thus $hgh^{-1} = ghh^{-1} = ge = g$. $\qquad\square$

That is, for any cyclic group, the number of conjugacy classes is always $N = |G|$ with $C_i = 1$ for all $i = 1, \dots, N$. Note that the total number of conjugacy classes of a group does not depend on the number of divisors of the order of the group. For example, for the cyclic group C_9 of order $|G| = 9$, there are 9 conjugacy classes but only 3 divisors, $1, 3$ and 9.

EXAMPLE 2.30 Conjugacy classes of the cyclic group C_3 (rotational symmetries of the equilateral triangle): Since C_3 is abelian and of order 3, there are three classes: $\{e\}$, $\{R_1\}$ and $\{R_2\}$.

A straightforward consequence of the partition into disjoint union of conjugacy classes is the following:

THEOREM 2.31 *Let $|C_i|$ denote the cardinality of (i.e., number of elements in) a conjugacy class C_i, $i = 1, \dots, N$ and N the total number of conjugacy classes. Since each element of G is in one and only one conjugacy class, we have that*

$$|G| = \sum_i |C_i|,$$

where i ranges over the number of distinct conjugacy classes of G.

Proof Follows from Theorem 2.28. $\qquad\square$

We will see that the conjugacy class structure of a group has important consequences for the representation theory of groups.

2.6 Normal Subgroups and Homomorphisms

Another important type of equivalence relation is given by the *cosets*. Let G be a group and H a subgroup of G. A *right coset* of H in G is a set of the form

$$Hg = \{hg : h \in H\}$$

for some fixed $g \in G$, while a left coset is a set of the form

$$gH = \{gh : h \in H\}.$$

Right and left cosets are equivalence classes of the equivalence relations $\sim_r$ and $\sim_l$ respectively, where

$$a \sim_r b \Leftrightarrow (\exists h \in H)(ha = b),$$
$$a \sim_l b \Leftrightarrow (\exists h \in H)(ah = b).$$

PROPOSITION 2.32 *Let H be a subgroup of the group G.*

  (i) *Any right or left coset of H has the same number of elements of H.*
 (ii) *Any two distinct left cosets of H are disjoint, and similarly for right cosets.*
(iii) *If G is finite, then the size of H divides the size of G; the quotient is the number of (left or right) cosets. This quotient is called the index of the subgroup H in G.*

The last part is the celebrated *Lagrange Theorem*.

The subgroup N of G is *normal* if the left and right cosets coincide: $Ng = gN$ for all $g \in G$. If this is the case, then we can define a "multiplication" on the set of (left or right) cosets by the rule

$$(Ng_1)(Ng_2) = N(g_1 g_2).$$

It can be shown that this operation is well-defined (that is, independent of the choices of g_1 and g_2 to represent the cosets), and makes the set of cosets into a group. This group is called the *quotient group* or *factor group* of G by N, written G/N.

Clearly, if G is an abelian group, then every subgroup of G is normal. But other groups also have this property, for example, the quaternion group (see the Exercises).

Thus the presence of a normal subgroup enables us to break the group G down into two smaller groups N and G/N.

Normal subgroups and quotient groups have a natural connection with homomorphisms. If $\phi: G \to H$ is a homomorphism, then the *image* of ϕ is the subset $\phi(G) = \{\phi(g) : g \in G\}$ of H, while the *kernel* of ϕ is the set $\ker(\phi) = \{g \in G : \phi(g) = 1\}$.

PROPOSITION 2.33   *Let $\phi : G \to H$ be a homomorphism. Then*

(i) *$\phi(G)$ is a subgroup of H;*
(ii) *$\ker(\phi)$ is a normal subgroup of G;*
(iii) *$G/\ker(\phi)$ is isomorphic to $\phi(G)$.*

PROPOSITION 2.34

If N is a normal subgroup of G, then the map $g \mapsto Ng$ is a homomorphism from G to G/N with image G/N and kernel N.

We mention here one instance which will be important in the sequel. The group G is a *double cover* of the group H if G has a normal subgroup Z of order 2 such that G/Z is isomorphic to H.

2.7 Representations

We have seen that matrix groups can be convenient to work with. We will also see abstract groups in terms of generators and relations, or multivector groups in Clifford algebras; we have seen homomorphisms between groups and isomorphisms between groups that are "practically the same". Matrix groups can be particularly convenient, and mappings from groups to matrix groups are clearly of particular interest, which motivates the following definition.

DEFINITION 2.35 (Representation)   A homomorphism of a group G onto a group of non-singular $n \times n$ matrices Γ_g, $g \in G$, is said to "represent" an abstract group as a matrix group. The group of matrices Γ_g is called an n-dimensional representation Γ of G.

Faithful: A representation for which this mapping is in fact a bijection is called a *faithful* representation,

Unfaithful: and one where it is a homomorphism but not an isomorphism is called *unfaithful.*

Trivial: Mapping all elements to the identity is a representation, which is called the *identical/trivial representation.*

Regular: The representation of a group G by $|G| \times |G|$ permutation matrices (defined by right multiplication) is called the *regular representation.*

We say a bit more about the regular representation. The *Cayley table* of a finite group $G = \{g_1, \ldots, g_n\}$ is the $n \times n$ array whose rows and columns are indexed by $\{1, \ldots, n\}$, with the element in row i and j being k if $g_i g_j = g_k$. Now column j represents a permutation of $\{1, \ldots, n\}$, carrying i to k if $g_i g_j = g_k$. The map taking the element g_j to this permutation (for $j = 1, \ldots, n$) is a faithful homomorphism from G into the symmetric group $\mathfrak{S}_n$. Hence we have *Cayley's theorem*:

THEOREM 2.36   *Every group of order n can be embedded as a subgroup of the symmetric group $\mathfrak{S}_n$.*

EXAMPLE 2.37   The 4×4 matrix version of the quaternion group from Example 2.12 is a faithful representation of the more abstract quaternion group given in terms of generators and relations, whilst the 3×3 matrices from Example 2.14 describing 180-degree rotations in the principal coordinate planes is also a representation, but an unfaithful one (since the quaternions are a double cover).

A basis transformation in a vector space "jumbles up" components of vectors and matrices. Invariant quantities are therefore of particular interest. Any invertible square matrix defines a basis transformation, which is a similarity transformation:

DEFINITION 2.38 (Similarity, similarity transformations and equivalent representations)   An invertible matrix S defines a mapping of the matrix A to the matrix A' as

$$A' = SAS^{-1}.$$

This mapping is called a *similarity transform*, and A is said to be *similar* or *equivalent* to A'. Representations related by similarity are called *equivalent*, i.e., an equivalent representation Γ' of G to a representation Γ of G satisfies $\Gamma'_g = S\Gamma_g S^{-1}$ for every $g \in G$ for some similarity transformation determined by an invertible matrix S.

PROPOSITION 2.39   *Similarity is an equivalence relation on the space of square matrices.*

Proof   An exercise: follow the proof in Example 2.22. $\square$

Quantities that are invariant under such similarity transformations, i.e., that do not depend on the choice of basis for a matrix representation, are therefore of particular interest! Determinant, trace and order of a group element are easily seen to be invariant under similarity transformations. This motivates the following definitions:

DEFINITION 2.40 (Characters and character systems)   Let Γ be a d-dimensional representation of a group G. Then the *character*, χ_g, of a group element $g \in G$ is simply the trace

$$\chi_g = \mathrm{Tr}\left(\Gamma_g\right).$$

The set of all $|G|$ characters is called the *character system* of the representation Γ of G.

THEOREM 2.41 (Characters as invariants) *All the members of a conjugacy class of a group G have the same character in a given representation Γ. Similarly, elements in equivalent representations have the same character.*

Proof If elements g and g' of a group G are in the same conjugacy class then by definition $g' = hgh^{-1}$ for some element $h \in G$. Hence

$$\Gamma_{g'} = \Gamma_{hgh^{-1}} = \Gamma_h \Gamma_g \Gamma_{h^{-1}} = \Gamma_h \Gamma_g \Gamma_h^{-1}$$

by the homomorphism properties of representations. Thus

$$\chi_{g'} = \mathrm{Tr}(\Gamma_{g'}) = \mathrm{Tr}(\Gamma_h \Gamma_g \Gamma_h^{-1}) = \mathrm{Tr}(\Gamma_g \Gamma_h^{-1} \Gamma_h) = \mathrm{Tr}(\Gamma_g) = \chi_g$$

by the cyclic property of the trace.

Similarly, equivalent representations are related by similarity, i.e., by definition $\Gamma'_g = S\Gamma_g S^{-1}$, and thus for any $g \in G$ we have

$$\chi'_g = \mathrm{Tr}(\Gamma'_g) = \mathrm{Tr}(S\Gamma_g S^{-1}) = \mathrm{Tr}(\Gamma_g S^{-1} S) = \chi_g.$$

$\square$

Above we have seen how to construct larger groups from smaller building blocks, e.g., via the (semi-)direct product. The simple groups are the analogue of the prime numbers: the irreducible building blocks of group theory.

Similarly, given two representations one can construct a larger representation simply by taking the direct sum, i.e., by having block-diagonal matrices where the small representations sit in the individual blocks $\Gamma' = \Gamma^1 \oplus \Gamma^2$. Of course, such a block-diagonal form may not be obvious in an equivalent representation, i.e., once a similarity transformation has masked this block-diagonal form! Similarly and conversely, one can thus wonder what the prime – the "irreducible" – building blocks of representations of a group G are. This motivates the following (slightly more general) definition:

DEFINITION 2.42 An $(m + n)$-dimensional representation Γ is *reducible* if there exists an equivalent representation Γ' such that

$$S\Gamma_g S^{-1} = \Gamma'_g = \begin{pmatrix} \Gamma'_{g,11} & \Gamma'_{g,12} \\ 0 & \Gamma'_{g,22} \end{pmatrix} \ \forall \ g \in G,$$

where $\Gamma'_{g,11}$ and $\Gamma'_{g,22}$ are $m \times m$ and $n \times n$ matrices, respectively, and *irreducible* otherwise. It is said to be *completely reducible* if $\Gamma'_{g,12} = 0$, i.e., if it can be fully block-diagonalised

$$S\Gamma_g S^{-1} = \Gamma_g' = \begin{pmatrix} \Gamma_{g,11}' & 0 \\ 0 & \Gamma_{g,22}' \end{pmatrix} \ \forall \ g \in G.$$

The $(m+n)$-dimensional matrices are the *direct sum* of the m- and n-dimensional matrices, which can be written as

$$\Gamma_g' = \Gamma_g^1 \oplus \Gamma_g^2,$$

or more generally

$$\Gamma_g' = \bigoplus_{i=1}^{n} \Gamma_g^i,$$

for (not necessarily distinct) irreducible representations ('irreps') Γ_g^i.

Irreducibility places exceedingly strong conditions on the entries that matrices in representations can have. The result is beautifully simple, but we shall omit the details of the argument here, as they are quite technical – as signaled by two lemmata by Schur, Maschke's theorem, defining a "group invariant" inner product

$$\{x, y\} := \frac{1}{|G|} \sum_{g \in G} \left(\Gamma_g x, \Gamma_g y \right) \tag{2.7}$$

with respect to which the matrices Γ_g in our representation Γ are unitary, and then using the eigendecomposition of Hermitian matrices. Since eigenvectors of Hermitian matrices are orthogonal, it is not surprising that the result relates to a statement on orthogonality:

THEOREM 2.43 (Grand Orthogonality Theorem)   *Let Γ^k and $\Gamma^{k'}$ be inequivalent irreducible representations of a group G with dimensions d and d', respectively. Then*

$$\sum_g (\Gamma_g^k)_{ij} (\Gamma_g^{k'})_{i'j'}^* = \frac{|G|}{d} \delta_{ii'} \delta_{jj'} \delta_{kk'}.$$

This profound theorem is thus a three-way orthogonality theorem: the right-hand side is zero in three cases – when the representations are inequivalent, and when the rows or columns of the respective representation matrices are different. We will largely use this result in the simpler form where we have taken the trace and work at the level of orthogonality of character vectors (see below). But broadly, we can think of the matrix entries of irreducible representations Γ^k as populating $|G|$-dimensional vectors

$$V_{ij}^k = \left((\Gamma_1^k)_{ij}, (\Gamma_2^k)_{ij}, \ldots, (\Gamma_{|G|}^k)_{ij} \right).$$

As said above, by the Grand Orthogonality Theorem two such vectors are orthogonal if they differ in any of the indices i, j or k, since with our definitions and the standard complex inner product

$$\sum_g (\Gamma_g^k)_{ij}(\Gamma_g^{k'})_{i'j'}^* = \left(V_{ij}^k, V_{i'j'}^{k'}\right) = \frac{|G|}{d}\delta_{ii'}\delta_{jj'}\delta_{kk'}.$$

The $|G|$ on the right-hand side essentially comes from summing over the regular representation , which decomposes as the sum over all the irreducible representations! In a $|G|$-dimensional space, there are at most $|G|$ mutually orthogonal vectors. Let $d_1, d_2, \ldots, d_k$ denote the dimensions of the irreducible representations. Then for each representation with dimension d_i, the entries of the $d_i \times d_i$ matrices lead to d_i^2 mutually orthogonal vectors. Summing over the regular representation, i.e., over all irreducible representations thus yields

$$\sum_i d_i^2 = |G|.$$

Thus the squared dimensionalities of the irreducible representations total the order of the group, i.e., one needs a set of integers whose squares sum to the order of the group. Of course the trivial representation is always one of the irreducible representations so one of these integers is always 1, but the other integers – and how many of them – are trickier. A theorem below comes to the rescue (at least, partially) in that the number of these irreps is given by the number of conjugacy classes. In order to prove this, we effectively take the trace of the Grand Orthogonality Theorem and derive a similar character orthogonality relation:

THEOREM 2.44 (Character Orthogonality Theorem)   *Consider the inequivalent irreducible representations Γ^k and $\Gamma^{k'}$ of a group G, and in particular their character systems χ_g^k and $\chi_g^{k'}$ (with $g \in G$). Then one has the orthogonality relationship*

$$\left(\chi_g^k, \chi_g^{k'}\right) = \sum_g \chi_g^k(\chi_g^{k'})^* = |G|\delta_{kk'}.$$

Proof   Starting with the Grand Orthogonality Theorem $\sum_g(\Gamma_g^k)_{ij}(\Gamma_g^{k'})_{i'j'}^* = \frac{|G|}{d}\delta_{ii'}\delta_{jj'}\delta_{kk'}$ and taking traces $j = i$ and $j' = i'$ one gets

$$\sum_g (\Gamma_g^k)_{ii}(\Gamma_g^{k'})_{i'i'}^* = \frac{|G|}{d}\delta_{ii'}\delta_{kk'},$$

then summing over i and i' on the left-hand side produces

$$\sum_g \sum_i \sum_{i'} (\Gamma_g^k)_{ii}(\Gamma_g^{k'})_{i'i'}^* = \sum_g \mathrm{tr}(\Gamma_g^k)\mathrm{tr}(\Gamma_g^{k'})^* = \sum_g \chi_g^k\chi_g^{k'*} = \left(\chi_g^k, \chi_g^{k'}\right). \quad (2.8)$$

Whilst summing over i and i' on the right-hand side, we get

$$\frac{|G|}{d_k}\delta_{kk'}\sum_i\sum_{i'}\delta_{ii'} = \frac{|G|}{d_k}\delta_{kk'}\underbrace{\sum_i 1}_{d_k} = |G|\delta_{kk'}.$$

This results in

$$\left(\chi_g^k,\chi_g^{k'}\right) = \sum_g \chi_g^k\chi_g^{k'*} = |G|\delta_{kk'}.$$

$\square$

The character version of the orthogonality theorem effectively shows that character systems of different irreducible representations are orthogonal. We hinted above that the number of irreps is given via the following theorem:

THEOREM 2.45 (Number of irreducible representations)   *The number of inequivalent irreducible representations is equal to the number of conjugacy classes of the group G.*

Proof   We proved in Theorem 2.41 that characters are invariants within a given conjugacy class C_i. Denote by N the number of conjugacy classes of a group G and by $|C_i|$ their cardinality. Then the character system for an irrep is simply given by the N class invariants χ_i^k. Summing over conjugacy classes, the character orthogonality formula can be restated as

$$\sum_{i=1}^{N}|C_i|\chi_i^k\chi_i^{k'*} = |G|\delta_{kk'},$$

where summation is now only over different conjugacy classes rather than over all elements, as the cardinalities $|C_i|$ take into account the multiplicities.

Consider the class character sets χ_i^k and $\chi_i^{k'}$ belonging to irreps Γ^k and $\Gamma^{k'}$ of the group G. We define the rescaled characters $X_i^k \equiv \sqrt{|C_i|}\chi_i^k$ as well as row vectors containing these rescaled characters:

$$V^k = (X_1^k,X_2^k,\ldots,X_N^k) \text{ and } V^{k'} = (X_1^{k'},X_2^{k'},\ldots,X_N^{k'}).$$

We can thus view the character orthogonality theorem as a statement of orthogonality of these N-vectors. There can be at most N mutually orthogonal such vectors, and in order to span the whole space, there need to be exactly N such vectors and thus irreducible representations. $\square$

We can therefore write the following square ($N \times N$) grid of characters of irreducible representations: firstly, the $|G|$ group elements can be grouped together as N conjugacy classes as long as orthogonality is with respect to a scalar product that takes into account the correct multiplicities. Secondly, in order to span that space there need to be N irreducible representations whose character systems we can write as the above row vectors. These square grids are called *character tables*. We will thus use the Character Orthogonality Theorem to create such character tables below. For now, let us look at characters for conjugacy classes of the rotational tetrahedral group considered in Examples 2.13 and 2.27.

EXAMPLE 2.46 (Characters for the rotational tetrahedral group in the faithful 3×3 rotation representation)   Characters give a clue as to the geometric interpretation of a group element. For instance, the identity matrix is the only one with trace 3. Three matrices have trace -1, which turns out are just the 180-degree rotations. The 120-degree rotations all have trace (and thus character) 0 (both conjugacy classes).

EXAMPLE 2.47   For the group $\mathfrak{S}_3$, we have $|G| = 6$, and we have seen in Example 2.25 that there are three conjugacy classes: the identity, the two rotations and the three reflections. Thus, we also have three irreducible representations of some dimensions d_i (of which one is the trivial representation with $d = 1$) such that the squares sum to $|G| = 6$. The only solution is

$$\sum_k d_k^2 = 1^2 + 1^2 + 2^2 = 6.$$

Thus there is the trivial representation of dimension 1 we already know about, and another one-dimensional irrep, as well as one of dimension two. The missing one-dimensional irrep turns out to be parity, or the determinant of the identity and two rotations $(+1)$ and the three reflections (-1). The two-dimensional representation is the faithful rotation/reflection representation in 2D acting on an equilateral triangle, as seen in Figure 2.2. From such an explicit set of 2×2 matrices acting on the vertices, one could read off the characters for the identity (2 of course, as it is two-dimensional), the rotations (-1) and reflections (0).

To summarise, the three main relations regarding the number of irreducible representations of a finite group G and their dimensions are thus

(i) The number of irreps is equal to the number of conjugacy classes.

(ii) The squared dimensions give the order of the group: $\sum_k d_k^2 = |G|$.

(iii) Character vectors are orthogonal: $\left(\chi_g^k, \chi_g^{k'} \right) = |G| \delta_{kk'}$.

We will encounter these again more concretely in the context of the representation theory of the polyhedral groups and the McKay correspondence in

Section 4.5. We now give some examples of finding just the dimensions before discussing whole character tables.

EXAMPLE 2.48 (Number of irreps for the quaternion group)   As we have seen, there are 5 conjugacy classes in total, partitioning the group. The order is $|Q| = 8$ and since $\sum d_i^2 = |Q|$ the irreducible representations can at most be of dimension two. In fact, since there are 5 of them (because of the 5 conjugacy classes), the only solution is $(1, 1', 1'', 1''', 2)$.

EXAMPLE 2.49 (Number of irreps for the rotational and binary tetrahedral group)   The tetrahedral group has 4 conjugacy classes and its order is 12. Similarly to the quaternion group, we therefore need to have irreps of dimensions $(1, 1', 1'', 3)$. It is interesting to note here (preempting Section 4.5) that the binary tetrahedral group of order 24 has three further conjugacy classes to account for the additional 12 elements, which leads to three 2-dimensional irreps, bringing the total for the binary tetrahedral group to $(1, 1', 1'', 2, 2', 2'', 3)$.

EXAMPLE 2.50 (Number of irreps for the rotational and binary icosahedral group)   The icosahedral group has 5 conjugacy classes and order 60. The only solution is $(1, 3, 3'', 4, 5)$. Similarly to above, going to the binary group doubles the order to 120 and introduces a further 4 conjugacy classes, with $(2, 2', 4', 6)$ accounting for the additional 60 group elements. The total set of $(1, 2, 2', 3, 3'', 4, 4', 5, 6)$ is quite iconic and will make many mysterious appearances throughout this book!

Having shown how the characters of a representation work and how the numbers of irreps, conjugacy classes and dimensions of the irreps interrelate, we now discuss a whole character table, and in particular how orthogonality relations can be used to fill in missing entries (a little like a Sudoku!).

DEFINITION 2.51 (Character table)   As the number of conjugacy classes is the same as the number of irreducible representations, we can construct an $N \times N$ table of the character systems for a finite group G called the "character table".

EXAMPLE 2.52   We have seen earlier that in $\mathfrak{S}_3$ we have three conjugacy classes of identity, rotations and reflections (I, $\{R_i\}$ and $\{S_i\}$), and that thus we have two 1-dimensional irreducible representations and one 2-dimensional. We construct the empty character table:

G	I	$\{R_i\}$	$\{S_i\}$
1			
$1'$			
2			

We can immediately fill in the first row and column: that the trivial representation assigns 1 and thus character 1 to every conjugacy class and that the character of the identity class in any dimension just gives the dimension:

G	I	$\{R_i\}$	$\{S_i\}$
1	1	1	1
$1'$	1		
2	2		

Now we need to use some of these orthogonality relations. The second row vector would have to be orthogonal to the first row vector (taking into account the multiplicities). But characters also need to satisfy the homomorphism property so effectively this representation needs to assign -1 to the reflections and $+1$ to the others:

G	I	$\{R_i\}$	$\{S_i\}$
1	1	1	1
$1'$	1	1	-1
2	2		

Now looking at the columns: we see that the first two columns determine the character of the 2D rotations as -1 by orthogonality. Since the first two columns only differ in the last row, the only way the reflection column can be blind to that difference is if that corresponding entry is 0. This fixes our completed character table as:

G	I	$\{R_i\}$	$\{S_i\}$
1	1	1	1
$1'$	1	1	-1
2	2	-1	0

Confirming the details of this is a good exercise.

EXAMPLE 2.53 (Character table for the Quaternion group) Let us find the character table of Q, with the given hints that both elements of the order two subgroup in Q (our $\mathbb{Z}/2\mathbb{Z}$ example from earlier) have the same character for the one-dimensional irreducible representations, and that the one-dimensional irreducible representations only have characters ± 1.

The order two subgroup can only be ± 1. So with the trivial information in the first row and column, this fixes also the second column entirely, as the -2 follows from orthogonality.

G	1	-1	$\pm i$	$\pm j$	$\pm k$
1	1	1	1	1	1
$1'$	1	1			
$1''$	1	1			
$1'''$	1	1			
2	2	-2			

Then there are 3 non-trivial one-dimensional irreps and the cyclic symmetry between $\{j, -j\}$, $\{i, -i\}$ and $\{k, -k\}$ conjugacy classes. So using the information that entries are only ± 1 it's easy to find that they are mostly -1 with one $+1$ in each row, with the cyclic symmetry accounting for $1'$, $1''$ and $1'''$. The 2-dimensional irrep has to be blind to the sign differences, so all other characters have to be zero ($2^2 + (-2)^2 = 8$ already anyway).

G	1	-1	$\pm i$	$\pm j$	$\pm k$
1	1	1	1	1	1
$1'$	1	1	1	-1	-1
$1''$	1	1	-1	1	-1
$1'''$	1	1	-1	-1	1
2	2	-2	0	0	0

The 4D representation above is $2 + 2$ in terms of irreps.

Character tables will make an appearance again in the context of the eponymous McKay correspondence in Section 4.5, where we will review the character tables of the binary polyhedral groups. Having reviewed some basics about algebras, groups, representations and invariants, as well as encountering some of the dramatis personae, we are now ready to encounter the first of our ADE sets, before exploring the exciting connections between them.

Exercises

2.1   Show that the number of ways of selecting k objects from a set of size n, where the order of the selection is not significant and selections may be repeated, is $\binom{n+k-1}{k}$. Hint: put $n-k+1$ boxes in a row; select $n-1$ of them and insert barriers. The multiplicity of the first, second, $\ldots$, nth object is the number of empty boxes before the first barrier, between the first and second barrier, ..., after the last barrier. Deduce that, if $\dim(V) = n$, then $\dim\left(\mathrm{Sym}^k V\right) = \binom{n+k-1}{k}$.

2.2 Let G be a group satisfying the presentation

$$G = \langle u, v, w, x, y : uv = w, vw = x, wx = y, xy = u, yu = v \rangle.$$

Prove that G is isomorphic to the cyclic group of order 11.

2.3 Prove that, for any two $n \times n$ matrices A and B, we have $\mathrm{Tr}(AB) = \mathrm{Tr}(BA)$. True or false: for any three $n \times n$ matrices A, B, C, we have $\mathrm{Tr}(ABC) = \mathrm{Tr}(CBA)$.

2.4 Show that the size of a conjugacy class of a finite group G divides the order of G. Deduce the theorem of Landau: given a positive integer N, there are only finitely many finite groups with N conjugacy classes. Find all groups with at most three conjugacy classes.

2.5 Show that the Cayley table of a finite group is a *Latin square*: that is, an $n \times n$ array containing n different letters such that each letter occurs once in each row and once in each column.

2.6 Let H be a subgroup of a group G. Show that the map $Hx \mapsto x^{-1}H$ is a bijection between right and left cosets of H. Deduce that, even if G is infinite, the numbers of right and left cosets of H are equal.

2.7 The *centre* of a group G is the set $Z = \{g \in G : gh = hg \text{ for all } h \in G\}$ of elements which commute with every element in G. Show that the centre of G is a normal subgroup of G and consists of all the elements of G whose conjugacy class has size 1.

2.8 The cyclic group C_3 has an automorphism τ mapping each element to its inverse. This generates a group of automorphisms of C_3 isomorphic to C_2. Show that the following groups are isomorphic:

- $\mathfrak{S}_3$;
- Dih_6;
- the semi-direct product of C_3 by C_2, where the map $\phi : C_2 \to \mathrm{Aut}(C_3)$ is an isomorphism.

2.9 - Show that any semi-direct product of G by H has a normal subgroup of G with quotient isomorphic to H.
- The converse is false; show that C_4 has a normal subgroup isomorphic to C_2 with quotient also isomorphic to C_2, but is not a semi-direct product of these two groups.

2.10 Let Γ be a representation of G. Show that the map $g \mapsto \det(\Gamma_g)$ is a 1-dimensional representation.

2.11 Let $\mathbb{F}$ be a field, and let $\tau_{a,b}$ be the map from $\mathbb{F}$ to itself given by

$$\tau_{a,b}(x) = ax + b.$$

Show that the set $G = \{\tau_{a,b} : a, b \in \mathbb{F}, a \neq 0\}$ is a group with the operation of composition. Show that the set $N = \{\tau_{1,b} : b \in \mathbb{F}\}$ is a normal subgroup of G isomorphic to the additive group of $\mathbb{F}$, and that G/N is isomorphic to the multiplicative group of $\mathbb{F}$.

2.12 Complete the details of filling in the character tables in Examples 2.52 and 2.53.

2.13 Explain the geometric meaning of each of the five conjugacy classes of the icosahedral group A_5. Generators of the icosahedral group A_5 are given in a certain three-dimensional irreducible representation as

$$D_3(g_2) = \frac{1}{2}\begin{pmatrix} -1 & \tau & \tau-1 \\ \tau & \tau-1 & 1 \\ \tau-1 & 1 & -\tau \end{pmatrix},$$

$$D_3(g_5) = \frac{1}{2}\begin{pmatrix} 1 & \tau & \tau-1 \\ -\tau & \tau-1 & 1 \\ \tau-1 & -1 & \tau \end{pmatrix}.$$

These generators are given in another, four-dimensional, irreducible representation as

$$D_4(g_2) = \frac{1}{4}\begin{pmatrix} -1 & -1 & -3 & -\sqrt{5} \\ -1 & 3 & 1 & -\sqrt{5} \\ -3 & 1 & -1 & \sqrt{5} \\ -\sqrt{5} & -\sqrt{5} & \sqrt{5} & -1 \end{pmatrix},$$

$$D_4(g_5) = \frac{1}{4}\begin{pmatrix} -1 & 1 & -3 & \sqrt{5} \\ -1 & -3 & 1 & \sqrt{5} \\ 3 & 1 & 1 & \sqrt{5} \\ \sqrt{5} & -\sqrt{5} & -\sqrt{5} & -1 \end{pmatrix}.$$

Calculate the characters of g_2, g_5 and g_2g_5 in both representations. Which of the five conjugacy classes can g_2g_5 be in? As usual, τ is the golden ratio and you can assume that it has the properties $\tau = \frac{1}{2}(1 + \sqrt{5}) = 2\cos\frac{\pi}{5}$ and $\tau^2 = \tau + 1$. Complete the character table of A_5.

2.14 Consider the group G generated by

$$I = \begin{pmatrix} 0 & 1 & 0 & 0 \\ -1 & 0 & 0 & 0 \\ 0 & 0 & 0 & 1 \\ 0 & 0 & -1 & 0 \end{pmatrix} \quad \text{and} \quad J = \begin{pmatrix} 0 & 0 & 1 & 0 \\ 0 & 0 & 0 & -1 \\ -1 & 0 & 0 & 0 \\ 0 & 1 & 0 & 0 \end{pmatrix},$$

and the group H consisting of

$$e = \begin{pmatrix} 1 & 0 & 0 \\ 0 & 1 & 0 \\ 0 & 0 & 1 \end{pmatrix}, \quad \tilde{i} = \begin{pmatrix} -1 & 0 & 0 \\ 0 & -1 & 0 \\ 0 & 0 & 1 \end{pmatrix},$$

$$\tilde{j} = \begin{pmatrix} -1 & 0 & 0 \\ 0 & 1 & 0 \\ 0 & 0 & -1 \end{pmatrix}, \quad \tilde{k} = \begin{pmatrix} 1 & 0 & 0 \\ 0 & -1 & 0 \\ 0 & 0 & -1 \end{pmatrix}.$$

Show that G and H are both representations of the quaternion group Q. Calculate the characters of G and H. Finally find the decomposition of both G and H in terms of irreducible representations of Q.

2.15 Let A be an abelian group of order n. Show that A has n irreducble characters, all of degree 1, and that these characters form a group (under multiplication) which is isomorphic to A.

2.16 The "displacement representation" is a representation of a symmetric physical object – for instance, 12 points denoting the vertices of an icosahedron in space – and is useful, e.g., for describing normal modes (vibrations) of such an object. The character for each conjugacy class is just given by how many vertices are invariant under each conjugacy class. So for instance, the identity class just counts the total number of vertices, and in the above example, the two classes of 5-fold rotations leave 2 antipodal points invariant whilst moving the other 10, and all other conjugacy classes have no fixed points. Calculate this representation as a sum of irreducible representations of the icosahedral group. You can also consider the decomposition for other icosahedral solids with vertices on symmetry axes and that for solids with no vertices on symmetry axes.

3

ADE Classifications

We describe the various seemingly disparate classification problems, of perhaps increasing sophistication. In the ensuing chapter, we will review how some of these are, in fact, interrelated, via ADE-correspondences. There are still many relations not yet clarified and it is expected that establishing their correspondences will lead to new mathematics. We will discuss more advanced ADE topics in the final chapter.

3.1 Polytopes

We first reintroduce the Platonic solids, but now in the language of polytopes which we introduced in Definition 2.3.

3.1.1 The Platonic Solids Revisited

In our present language, a Platonic solid is a regular polyhedron: a polytope in $\mathbb{R}^3$ with identical faces, each of which is a regular n-gon. The classification of these, part of which we already saw, is then that of (discrete, finite) rotational symmetry groups.[1] The regular n-gon obviously has the rotational symmetry of C_n. It also has the additional reflection symmetry of Dih_{2n}, because we can allow it to be reflected along any major diagonal. A prism has Dih_{2n} as its rotational symmetries, because of the C_n above along with an overall 2-fold rotation (a flip), and thus also the name "dihedral": a solid with two faces. We can thus view Dih_{2n} as a subgroup in either $O(2)$ or $SO(3)$.

---

[1] We will later see these as discrete finite subgroups of the continuous group $SO(3)$, which we will encounter later also in its guise as a Lie group.

40

Now, let us move on to the 5 Platonic solids. Because of the graph-duality (the tetrahedron is self-dual, the cube-octahedron pair, and the dodecahedron-icosahedron pair) there are only 3 different symmetries. It is well-known (q.v. [14]) that these are,[2] respectively, (i) the *tetrahedral* group $T \simeq \mathfrak{A}_4$, of size 12, (ii) the *octahedral* group $O \simeq \mathfrak{S}_4$, of size 24, and (iii) the *icosahedral* group $I \simeq \mathfrak{A}_5$, of size 60.

Therefore, the symmetries of the regular shapes in $\mathbb{R}^3$ fall into 2 infinite families, and 3 exceptional cases:

	Symmetry group G	$\lvert G \rvert$
n-gon	$\langle R \mid R^n = \mathbb{I} \rangle \simeq \mathbb{Z}/n\mathbb{Z}$	n
n-prisms	$\langle R, S \mid R^n = S^2 = (RS)^2 = \mathbb{I} \rangle \simeq \mathrm{Dih}_n$	$2n$
Tetrahedron	$\langle R, S, T \mid RST = R^2 = S^3 = T^3 = \mathbb{I} \rangle \simeq \mathfrak{A}_4$	12
Cube octahedron	$\langle R, S, T \mid RST = R^2 = S^3 = T^4 = \mathbb{I} \rangle \simeq \mathfrak{S}_4$	24
Dodecahedron icosahedron	$\langle R, S, T \mid RST = R^2 = S^3 = T^5 = \mathbb{I} \rangle \simeq \mathfrak{A}_5$	60

$$(3.1)$$

Voilà, our first ADE classification:

$$
\begin{array}{lcl}
\text{Infinite Family 1} & : & C_n \\
\text{Infinite Family 2} & : & \mathrm{Dih}_n \\
\text{3 Exceptionals} & : & \mathfrak{A}_4, \ \mathfrak{S}_4, \ \mathfrak{A}_5 \text{ also known as } T, O, I
\end{array}
$$

We note that the orders of rotational symmetries – given by the exponents in the relations for the group generators – of the n-gons are n, of the n-prisms $(2, 2, n)$ and for the Platonic solids $(2, 3, 3)$, $(2, 3, 4)$ and $(2, 3, 5)$, respectively. The Platonic triples are the 3 denominators of the Egyptian fractions in the Diophantine inequality (1.3) if we choose to rewrite it in the format

$$\frac{1}{p} + \frac{1}{q} + \frac{1}{r} > 1 \tag{3.2}$$

for $r = 2$ and $p, q > 2$. Note that, in fact, the n-prism triple $(2, 2, n)$ satisfies the inequality (3.2). These are not regular polyhedra but generalise the (degenerate) case of the regular n-gons (which give rise to the A-type family, the cyclic groups), to yield the D-type family (the dihedral groups), whilst the Platonic symmetries are effectively the E-type cases.

---

[2] Including the reflections for these Platonic solids yields groups twice as large, living inside $O(3)$, which are generally referred to in Coxeter nomenclature (see later in Section 3.2.1) as $A_3 \equiv \mathfrak{S}_4$ (not to be confused with the alternating groups), $B_3 = O \times \mathbb{Z}_2$ and $H_3 = I \times \mathbb{Z}_2$.

3.1.2　Tessellations of the Plane

The polytopes above can also be viewed as regular tessellations of the sphere (positive curvature). Since the sphere is compact, these tilings are necessarily finite. For example, the icosahedron, octahedron and tetrahedron are different tessellations of the sphere with triangles. Triangles are tiles around the 3-fold symmetry axes, which all three symmetries possess. Other tilings can occur with respect to other symmetry axes; e.g., the dodecahedron is a tiling with pentagons centred around the 5-fold axes of icosahedral symmetry.[3] The allowed tessellations of the sphere have symmetry orders given by the above triples $(2,3,3)$, $(2,3,4)$ and $(2,3,5)$. There is already an interesting parallel emerging with the (E_6, E_7, E_8) diagrams (count vertices on the three arms of the diagrams), but we will defer discussion of correspondences between ADE sets until later.

One could therefore also contemplate regular tessellations of the plane. This will necessarily lead to infinite tilings and lattices, as the plane is not compact like the sphere. The plane can be tiled with regular triangles, squares and hexagons, corresponding to, e.g., the triangular, square and hexagonal lattice. It is also possible for these tilings to be chiral, i.e., not have reflection symmetry. The crystallographic restriction theorem states 5-fold symmetry and n-fold symmetry for $n = 7$ and above are incompatible with a lattice structure. It states that only 2-, 3-, 4- and 6-fold symmetry is compatible with lattices.

The allowed 2D lattices/tessellations have symmetry axes with rotational symmetry orders given by the triples $(3,3,3)$, $(2,4,4)$, $(2,3,6)$ [15]. In terms of the Egyptian fraction problem from the angles meeting at each vertex, for the plane the analogous equation is the following equality:

$$\frac{1}{p} + \frac{1}{q} + \frac{1}{r} = 1 \tag{3.3}$$

where p, q, r are positive integers. We can thus appreciate that these triples are again the solutions to the corresponding Egyptian fraction problem. We note here that they are related to the above triple by simply increasing one of the integers by one. These triples also again have a tantalising connection with the extended $(\tilde{E}_6, \tilde{E}_7, \tilde{E}_8)$ diagrams. We will discuss the links between ADE and affine ADE in various contexts such as Sections 3.3 and 4.3.

There are also tilings of the hyperbolic plane, but they won't concern us here. This division into three cases, however, is a phenomenon that occurs

---

[3] An icosahedral tiling with tiles centred around the 2-fold axes are also possible. One is given by the rhombic triacontahedron, with a tile shape of a golden rhombus. This is not a regular n-gon and the rhombic triacontahedron is a Catalan solid. In fact it is dual to the icosidodecahedron, which is a triangle-pentagon tiling, and thus an Archimedean solid. We will encounter the icosidodecahedron again as the root system of H_3: its vertices lie exactly on all the 2-fold axes.

more widely throughout mathematics, e.g., that curvature can be positive, zero, or negative (corresponding to sphere, plane, hyperbolic space or S, E, H). A similar classification between spherical, Euclidean and hyperbolic occurs for 2D Coxeter groups (see Section 3.2.2), which is determined by whether the determinant of the Cartan matrix is positive, zero, or negative.

3.1.3 The Exceptionals and Trinities

In our first ADE example we have already seen that sometimes the three exceptional cases (here the symmetries of the Platonic solids) share some similarities amongst themselves but seem quite different from the two infinite families (symmetries of the n-gons and prisms).

Indeed, Arnold [7] has also considered these three exceptionals on their own, which he termed "Trinities". Like ADE classifications, such sets of three exceptional cases turn up throughout mathematics. Of course, each ADE set has such exceptional cases, but there are also stand-alone "Trinities" where 2 infinite families do not exist or are not obvious. We have already seen some examples as the symmetry groups of the Platonic solids (T, O, I) as well as the E-type diagrams (E_6, E_7, E_8). Fundamentally, his idea was to relate all such trinities to the three normed division algebras of real numbers, complex numbers and the quaternions

$$(\mathbb{R}, \mathbb{C}, \mathbb{H}). \tag{3.4}$$

That trinities of exceptionals should relate to this perhaps the most fundamental trinity of division algebras (3.4) already suggests something profound. A total of 25 trinities were observed in [2], ranging across the entire spectrum of mathematics. We will leave a fuller discussion of Arnold's trinities to the concluding chapter. He thought of them as webs of connections across different areas of mathematics. Some connections are more immediate and obvious or at least plausible, whilst others are mysterious and hint at a deep, as-yet-not-understood connection. In fact, using such conjectures about connections between different areas fuelled his creativity, as did thinking of examples of a "real" theory and attempting to "complexify" or "quaternionify" it. Such an approach contributed to his highly creative and original insights into many different areas of mathematics.

The web of ADE connections and the creative search for ADE patterns is very similar in many ways. As we proceed in this book, we will also suggest additional trinities such as $((2, 3, 3), (2, 3, 4), (2, 3, 5)), (2T, 2O, 2I), (12, 18, 30)$ and $(24, 48, 120)$, as well as J. McKay's old trio

$$(\text{Monster}, \ \text{Baby Monster}, \ \text{Fischer's Group}),$$

which turn up in different guises and hint at an extension to a full ADE correspondences.

A good way of thinking about Trinities is perhaps that it is easy to find three
exceptional examples in different areas of mathematics, which can serve as a
bootstrap to uncover the fuller ADE sets and correspondence. As an anecdote,
in the 1970s, Jaap Seidel and Jean-Marie Goethals were working on graphs
with the smallest eigenvalue -2. Both reported that they had found one infinite
family and three sporadic cases. Upon comparing notes with one of the authors
(PC), it turned out they had found the same exceptional cases but different
infinite families, completing the picture to a full ADE pattern with *two* infinite
families. Later, we will give a similar example of a trinity giving an idea as to
which infinite families could complete a correspondence to a full ADE pattern.

Our hope with this book is that the many tantalising connections between
different areas of mathematics inspire the next generation of mathematicians
and lead to the formulation of new conjectures and the development of interesting mathematics that bridges the divides between these different areas of
mathematics as we currently understand them.

3.2 Root Systems and Polyhedral Groups

Back to the symmetry groups of the Platonic solids. The symmetries we considered above are the rotational symmetries, which are subgroups of SO(3).
But these solids also have reflection symmetries. The symmetry groups of rotations and reflections are contained in O(3), and have twice the size of the rotation groups, containing the latter as subgroups of index 2. We will encounter
a non-trivial double cover in Spin(3) = SU(2) soon (Section 3.4). But for now
we focus on reflections, and the symmetry groups they generate.

We will examine them in a systematic manner and study them using root
systems. While this subject may seem advanced, it involves no more than linear algebra, and really ought to be taught at an early undergraduate level, rather
than leaving it, as is customarily done, to a post-graduate course on Lie algebras (to which we will turn much later, in Section 3.5).

First, we recall, from elementary geometry of a vector space, the **reflection**
of a vector x in a hypersurface defined by its unit normal vector n (Figure 3.1).
We decompose $x = x_\parallel + x_\perp$ where $x_\parallel$ is the component of x parallel to n, and $x_\perp$
is the orthogonal component. The reflection[4] amounts to a reversal of $x_\parallel$ such

---

[4] The last equality holds when there is also an inner product on the vector space such that one
can express the parallel component through the inner product.

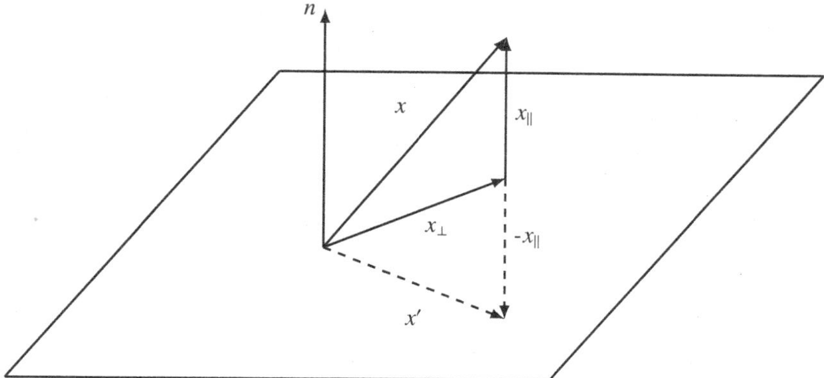

Figure 3.1 Reflection of a vector x in the hyperplane defined by the unit normal vector n.

that

$$x \xrightarrow{\text{Reflection}} x' = x_\perp - x_\parallel = x - 2x_\parallel = x - 2(x \cdot n)n. \tag{3.5}$$

The backbone of reflection groups is given by algebraic objects called root systems. We will see later that some of them (the crystallographic root systems) in turn provide the backbone of Lie algebras (Section 3.5). So whilst they only involve some fairly intuitive geometry, they turn out to be rather fundamental.

3.2.1 Root Systems and Coxeter Groups

Prepared with the elementary result from (3.5), we now define[5]

DEFINITION 3.1 (Root system)   A *root system* is a collection Φ of non-zero vectors α (which are called "roots") spanning an n-dimensional Euclidean

[5] Other common definitions of root systems are more restrictive by also stipulating an axiom on the (half-)integrality of inner products of root vectors. This stems from the traditional route of starting with Lie algebras (which is the other way around from the one pursued here), which has additional constraints and only yields certain types of root systems by construction (namely the crystallographic ones).

vector space V endowed with a positive definite bilinear[6] form $(v, w) \mapsto (v \cdot w)$, which satisfies the two axioms:

(i) Φ only contains a root α and its negative, but no other scalar multiples: $\Phi \cap \mathbb{R}\alpha = \{-\alpha, \alpha\} \;\; \forall \; \alpha \in \Phi$.

(ii) Φ is invariant under all reflections corresponding to root vectors in Φ: $s_\alpha \Phi = \Phi \;\; \forall \; \alpha \in \Phi$, where the reflection s_α in the hyperplane with normal α is given by

$$s_\alpha : \lambda \rightarrow s_\alpha(\lambda) = \lambda - 2\frac{(\lambda \cdot \alpha)}{(\alpha \cdot \alpha)}\alpha \; .$$

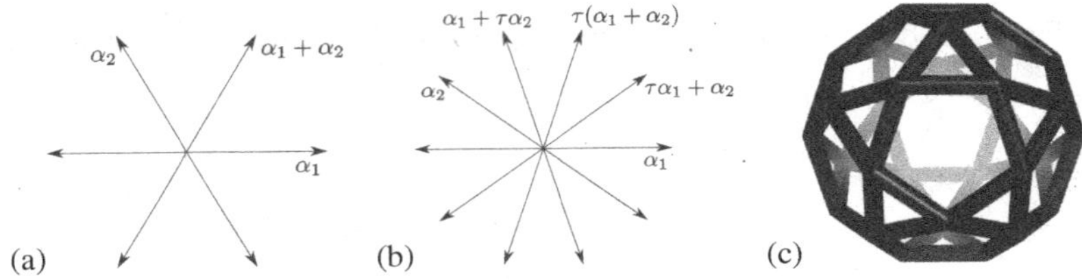

Figure 3.2 Examples of root systems (a) hexagonal $I_2(3 = A_2)$, (b) decagonal $I_2(5)$ and (c) H_3, the icosidodecahedron.

In Figure 3.2a and b, we show two examples of root systems, consisting respectively of 6 and 10 roots in $\mathbb{R}^2$. One can readily check that the two axioms are satisfied. The notation I_n, H_n, etc., may seem mysterious for now, but we will define them shortly in Section 3.2.2.

Remark: Root systems naturally give rise to a **reflection group**: each root vector defines a hyperplane (the one to which it is normal) and thus a reflection in that hyperplane. Each reflection is of course an involution (it is its own inverse). Multiplying together such reflections $s_i : x \rightarrow s_i(x) = x - 2\frac{(x|\alpha_i)}{(\alpha_i|\alpha_i)}\alpha_i$ generates a finite group. In general, multiplication of reflections of arbitrary collections of vectors would generate an infinite group. Root systems are those special generating sets that are closed under reflections and generate a finite

---

[6] First, in the root system setting one only needs a vector space with an inner product (which is all one needs to construct the corresponding Clifford algebra, which we will introduce and discuss later). Second, Clifford algebras have a uniquely simple reflection formula. These two facts make Clifford algebras the ideal framework for reflection groups [16]. But since Clifford algebras actually also naturally provide spin double covers, we will defer discussion of Clifford algebras till we discuss the binary groups in Section 3.4.

group. This reflection group is in fact a **Coxeter group**, since the reflections s_i satisfy the following defining relations [17–20]:

DEFINITION 3.2 (Coxeter group)   A **Coxeter group** $C(s_i)$ is a group generated by a set of involutory (i.e., square to identity) generators $s_i, s_j \in S$ subject to relations of the form $(s_i s_j)^{m_{ij}} = \mathbb{I}$ with $m_{ij} = m_{ji} \geq 2$ for $i \neq j$. In other words,

$$C(s_i) := \left\langle s_i : (s_i s_j)^{m_{ij}} = \mathbb{I} \right\rangle, \quad m_{ij} = m_{ji} \begin{cases} \in \mathbb{Z}_{\geq 2} & i \neq j, \\ = 1 & i = j. \end{cases}$$

Remark: We now link our foregoing discussion together by noting that root systems can also be thought of as polyhedra, with the vectors pointing to the vertices. This is exemplified in Figure 3.2c, where the 3D polyhedron known as the icosidodecahedron corresponds to the 3D root system generating icosahedral symmetry, which we shall call H_3. It is defined by vertices lying on the 2-fold axes of icosahedral symmetry. (Vertices lying on the 5-fold axes give the icosahedron, whilst vertices on the 3-fold axes give the dodecahedron.) This icosidodecahedron is not Platonic, as it consists of triangles and pentagons, but Archimedean, and also has the icosahedral group as its symmetries.

Simple Roots: A subset Δ of Φ, called *simple roots* $\alpha_1, \ldots, \alpha_n$ is sufficient to express every element of Φ via linear combinations with coefficients of the same sign (which can be chosen by convention). This same sign condition allows one to define an ordering on the roots, with the *highest root* being the one with the highest total sum of coefficients. These simple roots are in a one-to-one correspondence with the simple reflections which generate the Coxeter group above.

The simple roots also allow us to distinguish two types of root systems:

(i) In a **crystallographic** root system, these linear combinations are $\mathbb{Z}$-linear combinations. For example, in Figure 3.2a, the root system $\Phi = I_2(3 = A_2)$, the simple roots are α_1 and α_2, which in $\mathbb{R}^2$ can be written explicitly as the vectors $\alpha_1 = e_1$ and $\alpha_2 = \cos(\frac{2\pi}{3})e_1 + \sin(\frac{2\pi}{3})e_2$. The other roots are $-\alpha_1, -\alpha_2$ and $\pm(\alpha_1 + \alpha_2)$.

(ii) For a **non-crystallographic** root system, one needs to go beyond integer coefficients in expanding an arbitrary root in terms of the simple ones. Typically, one needs to work in certain extended integer rings. For example, in Figure 3.2b, for the root system $\Phi = I_2(5) = H_2$, we have the simple roots $\alpha_1 = e_1$ and $\alpha_2 = \cos(\frac{4\pi}{5})e_1 + \sin(\frac{4\pi}{5})e_2$, and the other roots are $\alpha_1 + \tau\alpha_2, \tau\alpha_1 + \alpha_2$, and $\tau\alpha_1 + \tau\alpha_2$, together with their negatives.

Here, τ is the *golden ratio* $\tau = \frac{1}{2}(1 + \sqrt{5}) = 2\cos\frac{\pi}{5}$. The fancy way of saying this is that the coefficients of expansion live in the extended integer ring $\mathbb{Z}[\tau] = \{a + \tau b \mid a, b \in \mathbb{Z}\}$, with σ the Galois conjugate of τ as $\sigma = \frac{1}{2}(1 - \sqrt{5}) = -2\cos\frac{2\pi}{5}$ (i.e., τ and σ are the two solutions to the quadratic equation $x^2 = x + 1$).

This difference between crystallographic and non-crystallographic arises because some of these root systems are related to *lattices*. Lattices of course have integrality conditions – they can be achieved by translating the corresponding root system in integer steps along the symmetry axes such that the lattice extends to infinity. Those lattices derivable from root systems are then called **root lattices**. We will see the interplay with translations and the process of affinisation again later in Section 4.3.

Conversely, for non-integer linear combinations such as the ones based on the golden ratio above, there exists no associated lattice; hence they are non-crystallographic. Unrestricted translations with $\mathbb{Z}[\tau]$-coefficients would densely fill $\mathbb{R}^2$ since $\mathbb{Z}[\tau]$ is dense in $\mathbb{R}$. τ is of course closely related to 5-fold symmetry (cf. the icosahedron). From the crystallographic restriction theorem in 2D, it is well known that 5-fold symmetry is incompatible with a lattice structure. In fact, only 2-, 3-, 4- and 6-fold symmetry is compatible with lattices, in which case the inner products are half-integral.[7] These 2D lattices and the crystallorgraphic restriction theorem are of course related to the plane tessellations discussed in Section 3.1.2, themselves related to the affine ADE diagrams.

However, non-crystallographic root systems still define interesting reflection groups, even if they do not have an associated Lie algebra (which we will introduce later). In particular for icosahedral symmetry, this is the right approach – however, historically this Lie-centric approach has led to various connections being overlooked over decades by neglecting the importance of the non-crystallographic groups [21, 22].

Cartan Matrix: Since root systems are concerned with the relative orientations (and perhaps lengths) of the root vectors, finding a rotation-invariant way of capturing the essence of this configuration is useful. This is of course given by the inner products between simple roots:

DEFINITION 3.3 (Cartan matrix) The *Cartan matrix* of a set of simple roots $\alpha_i \in \Delta \subset \Phi$ is defined as the matrix

$$A_{ij} := 2\frac{(\alpha_i \cdot \alpha_j)}{(\alpha_j \cdot \alpha_j)}.$$

---

[7] We will see later in Section 3.5 that in Lie algebra structure theory one arrives at a root lattice – associated root systems are therefore always crystallographic in Lie theory.

For ADE cases, as we will shortly see, the roots are all of the same length ('simply-laced') such that the Cartan matrix is essentially the scalar products between simple roots and is thus symmetric. For different lengths the Cartan matrix acquires asymmetry by differential scaling by the lengths of the simple roots.

Note that we will be working with two different normalisations: first, roots of unit length $\alpha^2 = 1$, which is convenient in the Clifford algebra framework, and $\alpha^2 = 2$, which is standard in the Lie theory framework. We will discuss both concepts soon. Note that, in particular in the Lie theory context, it can be convenient to define the *coroot* $\alpha^\vee$ of a root vector α as $\alpha^\vee = 2\alpha/(\alpha \cdot \alpha)$. The normalisation with the factor of 2 used in Lie theory then simplifies various formulas. The basis dual to the simple roots is called the *weights* ω. The Cartan matrix essentially gives the expansion of the roots in terms of the weights $\alpha_i = \sum A_{ij}\omega_j$ (and vice versa for its inverse). The coefficients of the highest root with respect to the bases of simple (co)roots are called the (dual) *Coxeter labels*, respectively.

Coxeter–Dynkin Diagrams: A graphical representation of the geometric content of a root system is given by *Coxeter–Dynkin diagrams*, according to the following rule:

- draw one node for each simple root;
- orthogonal roots (i.e., $(\alpha_i|\alpha_j) = 0$) are not connected;
- roots at $\frac{\pi}{3}$ have a simple link;
- other angles $\frac{\pi}{m}$ have a link with a label m.

We will discuss graphs soon in Section 3.3. We give an important definition:

DEFINITION 3.4 A root system (and the corresponding Coxeter–Dynkin diagram) is called **simply-laced** if it consists only of simple links (there are no labels on them). That is, all roots are at 90° or 120° with each other.

For example, the root system H_3, in Figure 3.2, has one link labelled by 5 (via the above relation $\tau = 2\cos\frac{\pi}{5}$), as does its four-dimensional analogue H_4. This is of course related to the 5-fold symmetry. Thus, the Coxeter–Dynkin diagram for H_3 looks like

$$H_3: \quad \bullet \overset{5}{\rule{1cm}{0.4pt}} \bullet \rule{1cm}{0.4pt} \bullet \tag{3.6}$$

Thus, H_3 is not simply-laced.

On the other hand, $I_2(3)$ is a special case where the label is $n = 3$ (and thus suppressed) and the simple roots are at an angle $\frac{2\pi}{6}$. In fact, $I_2(3)$ is simply A_2.

In Lie theory, into which we shall delve in Section 3.5, again there is an additional constraint which ties the angle to the relative lengths, so that in Dynkin diagrams the length denotes the angle, whilst here the angle is denoted via the label. These two types of diagrams are of course closely related, at least for crystallographic root systems. By an abuse of notation it is also easy to use the names for the root system, reflection group, graph, Lie algebra and Lie group fairly inter-changeably. In Lie algebra theory simply-laced means that all simple roots have the same lengths. For root systems, it largely means that the diagram has no labels and thus that all the links are just a single line – there are only the implicit labels of 2 and 3, indicating that simple roots are either orthogonal or at 120 degrees.

3.2.2 Classification of Root Systems

Now that we have a concise way of illustrating root systems in diagrammatical form, we can list the different types. This gets us to the heart of the problem and the underlying theme of the book: classification patterns.

Without much further ado, we have the following classification theorem.

THEOREM 3.5 *There are 12 types of root systems Φ, with corresponding reflection group W, as follows:*

$$
\begin{array}{|c|c|c|}
\hline
\textit{Type} & |\Phi| & |W| \\
\hline
A_n & n(n+1) & (n+1)! \\
\hline
B/C_n & 2n^2 & 2^n n! \\
\hline
D_n & 2n(n-1) & 2^{n-1} n! \\
\hline
E_6 & 72 & 2^7 3^4 5 \\
\hline
E_7 & 126 & 2^{10} 3^4 5 \cdot 7 \\
\hline
E_8 & 240 & 12096 \cdot 240^2 \\
\hline
F_4 & 48 & \frac{1}{2} \cdot 48^2 \\
\hline
G_2 & 12 & 12 \\
\hline
H_2 & 10 & 10 \\
\hline
H_3 & 30 & 120 \\
\hline
H_4 & 120 & 120^2 = 14400 \\
\hline
I_2(n) & 2n & 2n \\
\hline
\end{array}
\tag{3.7}
$$

The corresponding Coxeter–Dynkin diagrams are drawn in Figure 3.3. The subscripts of the names give the rank, i.e., the number of simple roots / nodes.

The proof is of a graph-theoretic nature. It classifies graphs of positive type and uses the result that subgraphs of graphs of positive type are positive definite. We will see several simpler examples of this type of proof in Section 3.3 on graphs and we refer the interested reader to Section 2.7 in [23] for details of the proof.[8]

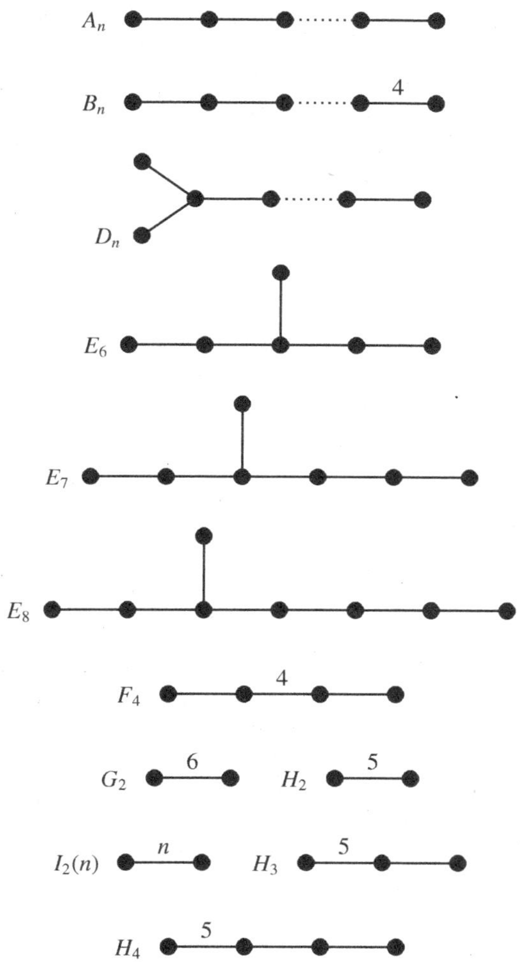

Figure 3.3 Classification of finite Coxeter groups and root systems.

We immediately see the emergence of the diagrams from Figure 1 at the start of the book. These correspond to the A, D and E-type root systems. The

[8] An argument by which the non-ADE crystallographic root systems are essentially built from the ADE root systems is contained in Section 5.2.

notation in the caption of Figure 3.2 now becomes clear. For instance, $I_2(3)$ has 2 simple roots, and hence 2 nodes, with the angle between them is $\frac{2\pi}{3}$ so the label of the link between the 2 nodes is marked 3. The reflection group is $\mathrm{Dih}_3 \simeq \mathfrak{S}_3$, of order 6.

Other familiar groups also feature. For example, the full tetrahedral group T_h is the reflection group for the A_3 root system. It is of order 24 and doubly covers the corresponding rotation group A_4 of order 12. The full octahedral group O_h is that of B_3; it is of order 48, doubly covering the rotational octahedral group S_4 of order 24. The full icosahedral group I_h is the reflection group for H_3; it is of order 120 and double covers $I \sim A_5$ of order 60. These are, as we recall, the symmetry groups of the Platonic solids. We will discuss double covers in detail later in Section 3.4.2.

ADE Root Systems: The higher-dimensional analogue of tetrahedral symmetry is a family called A_n with a graph consisting of a string of nodes (simply-laced). This[9] is the A in ADE. The generalisation of cubic/octahedral symmetry are two families called B_n and C_n that differ by their relative lengths, and are thus not simply-laced, but are also connected to another called D_n. This is simply-laced, and constitutes the D in ADE. The next are of course the 3 exceptional cases E_6, E_7, and E_8, which are again simply-laced, and complete our ADE set. There are another two exceptional root systems that are not simply-laced, called F_4 and G_2. These are all the crystallographic root systems, and can thus generate lattices. This result is of vital importance, so we summarise it as:

THEOREM 3.6 *The crystallographic root systems are $A_n, B_n, C_n, D_n, E_{6,7,8}, F_4$ and G_2, of which $A_n, D_n, E_{6,7,8}$ are simply-laced.*

We note that the lengths of the legs in the ADE diagrams are n for A_n, $(2, 2, n)$ for D_{n+2} and $(2, 3, 3)$, $(2, 3, 4)$ and $(2, 3, 5)$ for E_6, E_7 and E_8, respectively. Note the similarities with, and see Section 4.1 for further details on the connection with, the triples of rotational orders of the Platonic solids.

We remark that customarily, one calls the family

$$E_8, \ E_7, \ E_6, \ D_5, D_4, \ A_3, \ A_2, \ A_1$$

obtained by successive deletion of an appropriate single node, the **E-series**. Hence, $D_5, D_4, A_{3,2,1}$ are sometimes denoted as $E_{5,4,3,2,1}$. This is a special family with many nice properties, and even shows up in so-called grand-unified theories in physics (see, e.g., [24]).

---

[9] Not to be confused with alternating groups!

There are also non-crystallographic root systems that continue the alphabetical pattern (these are perhaps less familiar since they do not feature in Lie theory). The symmetries of the regular n-gons are given by the root systems $I_2(n)$ which are just the regular $2n$-gons. For example, the root system generating 5-fold rotations is a regular decagon (see Figure 3.2b)). In this terminology, it is also called H_2 in analogy with its 3D counterpart the icosahedral root system H_3 (which as a polyhedron is the icosidodecahedron in Figure 3.2c)) and its 4D analogue H_4 (aka the 600-cell). These are all the root systems for **finite Coxeter groups**, also known as Euclidean reflection groups. This is also reflected in the fact that the determinant of their Cartan matrices is positive, which is called the **spherical case**. Just to clarify the notation straight off, we have (in 2D)

$$\det(\text{Cartan Matrix}) \begin{cases} > 0 & \text{Spherical} & \text{finite Coxeter group,} \\ = 0 & \text{Affine or Euclidean} & \text{infinite Coxeter group,} \\ < 0 & \text{Hyperbolic} & \text{infinite Coxeter group.} \end{cases} \tag{3.8}$$

There are also Coxeter groups and root systems that are not finite and belong to the **Euclidean case**, i.e., the geometry of the plane. These are related to tilings/lattices as mentioned above in Section 3.1.2.

Explicit Representation of the ADE Root Systems Often, for calculation, it is useful to have an explicit list of the vectors in one of the ADE root systems. These can be given as follows. In each case, $\{e_1, e_2, \ldots, e_r\}$ is an orthonormal basis for the space in which the root system lives.

A_n: Take $r = n + 1$; then the vectors

$$\{\{e_i - e_j : 1 \leq i, j \leq r, \neq j\}$$

form a root system of type A_n. Note that these vectors are all orthogonal to the vector $e_1 + e_2 + \cdots + e_r$, and so they span a space of dimension $r - 1 = n$.

D_n: Take $r = n$. The vectors

$$\{\pm e_i \pm e_j : 1 \leq, j \leq n, i \neq j\}$$

form a root system of type D_n.

E_8: Take $r = 8$, and the vectors of the D_8 root system listed above together with the vectors

$$\left\{ \tfrac{1}{2}(\epsilon_1 e_1 + \cdots + \epsilon_8 e_8) : \epsilon_1, \ldots, \epsilon_8 \in \{+1, -1\}, \prod_{=1}^{8} \epsilon_i = +1 \right\}$$

form a root system of type E_8.

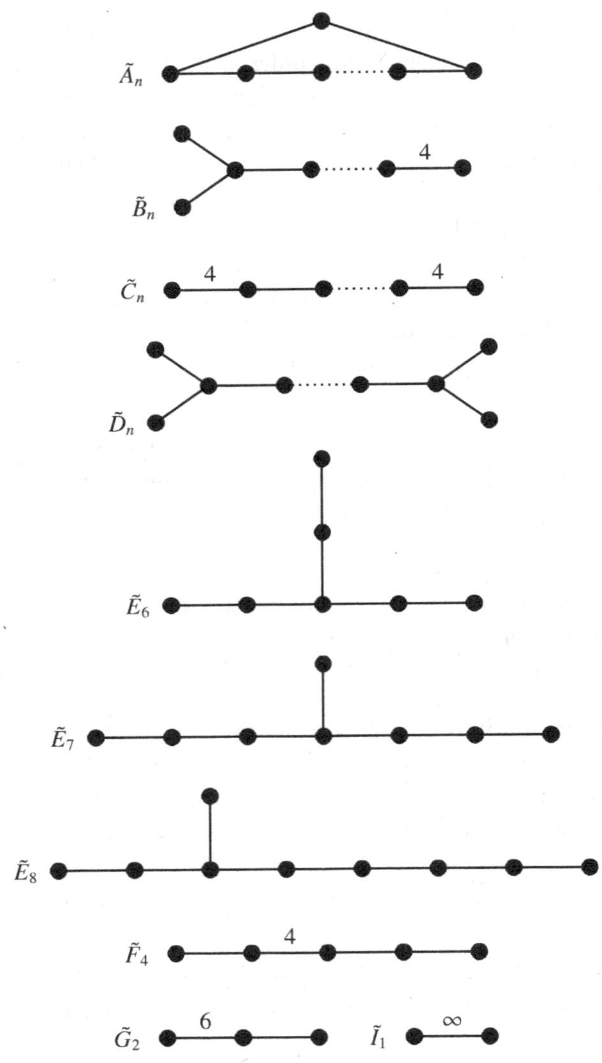

Figure 3.4 Classification of affine Coxeter groups and root systems.

E_7, E_6: The root system E_7 consists of the vectors of E_8 orthogonal to a fixed root, and E_6 consists of the vectors orthogonal to a pair of non-orthogonal roots.

Affine Root Systems Figure 3.4 shows these so-called affine cases, which are root lattices and understandably are connected with the crystallographic cases (non-crystallographic analogues would densely fill space upon adding a translation operator). These "in-between" cases have Cartan matrix with vanishing determinant. They can also be derived from the finite root systems via extending their root system via an affine root α_0, corresponding to adding an additional row and column in the Cartan matrix (so that the determinant is then zero). This is called a Kac–Moody[10] approach [28]. The resulting affine Coxeter groups are the Weyl groups occurring in the root lattices of Lie theory and this approach is also connected with the translations/affinisation process in Section 4.3.

Note that the affine root in each case can be written as a linear combination of the simple roots, e.g., $-\alpha_0 = 2\alpha_1 + 3\alpha_2 + 4\alpha_3 + 5\alpha_4 + 6\alpha_5 + 4\alpha_6 + 2\alpha_7 + 3\alpha_8$ for E_8 for some numbering of simple roots. Table 3.1 gives the integer coefficients of the highest/affine root that also gives the affine extension. We will encounter these again in a different guise in Sections 3.4.3, 4.3 and 4.5.

Type	Coefficients a_i
A_n	$1, 1, \ldots, 1$
B/C_n	$1, 2, \ldots, 2$
D_n	$1, 2, \ldots, 2, 1, 1$
E_6	$1, 2, 2, 3, 2, 1$
E_7	$2, 2, 3, 4, 3, 2, 1$
E_8	$2, 3, 4, 6, 5, 4, 3, 2$
F_4	$2, 3, 4, 2$
G_2	$3, 2$

Table 3.1 *The coefficients of expansion of the affine root α_0 in terms of the simple roots.*

Finally, there are the root systems with infinite Coxeter groups that have Cartan matrices with negative determinant, where roots can become imaginary. We will not be concerned with these, but the interested reader can refer to [20].

---

[10] A similar (but limited) approach was used first in [25] and then in [26] to define extended but finite structures with 5-fold symmetry that had some limited translational symmetry; a related construction combined the affine ADE affine roots with a projection from (A_4, D_6, E_8) to (H_2, H_3, H_4) (see Section 3.2.4) to achieve a similar, relaxed notion of an "affine extension" for non-crystallographic groups [27]. We note that the D_n family can be extended in three different ways in the Kac–Moody formalism and thus form three different lattices – however, only one of them is simply-laced. So the (simply-laced) ADE diagrams have a unique simply-laced affine version.

The classification problem is essentially a graph-theoretic one, which we will further discuss in Section 3.3.

3D Root Systems: The Platonic symmetries (T, O, I) (rotational) can now be written as (A_3, B_3, H_3) (reflection) in the language of root systems. Somewhat less obvious is that one can also have root systems consisting of (orthogonal) sums of "irreducible components" with the corresponding reflection groups being the direct product of the groups corresponding to the irreducible parts. For example, there is thus an infinite family of $A_1 \times I_2(n)$ groups and root systems in 3D.[11] We will show in Section 4.4 that, surprisingly, each 3D root system determines a root system in 4D. The traditional point of view would be that the lower-dimensional root systems can be obtained from the higher-dimensional ones by ignoring a dimension. This bottom-up construction however shows that these higher-dimensional cases can in fact be constructed from the lower-dimensional ones. The set of 3D root systems thus encompasses three exceptional cases and an infinite family. We will argue in Section 4.6 that one should perhaps also include the 2D root systems $I_2(n)$ in order to obtain a full ADE set.

4D Root Systems: The set of root systems in 4D consists of the usual A_4, B_4 and D_4 as well as the exceptional crystallographic case F_4 and the exceptional non-crystallographic root system H_4. In fact, D_4 is also in some sense exceptional, because in 4D the D_4 diagram acquires an additional symmetry called triality (by permuting the three legs in the diagram, see Figure 4.2 in Section 4.6), which is related to the corresponding Lie group $SO(8)$.[12] These three thus combine to form a Trinity (D_4, F_4, H_4) of exceptional objects in 4D – which Arnold had spotted somewhat randomly (see Section 4.4.2); but via the construction in Section 4.4 hinted at in the previous subsection, (D_4, F_4, H_4) can in fact be constructed from (A_3, B_3, H_3). This connection along with the link between (A_4, D_6, E_8) and (H_2, H_3, H_4) hints that there are some links crossing families and exceptional cases, as well as crystallographic and non-crystallographic ones.[13]

---

[11] Technically, one should distinguish between the direct sum of root systems $A_1 \oplus I_2(n)$ and the Coxeter group it generates, which is the direct product of groups $A_1 \times I_2(n)$. However, since we treat these fairly interchangeably anyway, we shall be forgiven such sloppy notation.

[12] This triality property is actually of surprising interest in superstring theory, which is a theory in 10 dimensions. $SO(8)$ describes the symmetries of the 8 transverse dimensions. Triality relates two spinor and one vector representation of $SO(8)$, which allows to prove the equivalence of the Green–Schwarz and Ramond–Neveu–Schwarz strings [29, 30].

[13] One of these links is the close connection between 3D geometry, Spin(3), SU(2), the quaternions and 4D (Section 3.4). All of this is we think best understood in terms of Clifford algebras, which we will discuss in due course (see Section 3.4.4). We argued in the last section

3.2.3 The ADE Root Systems

The crowning glory of the root system classification is of course the set of ADE root systems. We already introduced them but it is worth to devote a section to exalt their virtues. They are the simply-laced diagrams, and as we have seen, they also have uniquely corresponding simply-laced affine versions. They are crystallographic and thus can provide the backbone for Lie algebras and groups, to which we will turn in Section 3.5.

These are therefore the very reason why we talk about ADE sets, correspondences, and patterns, since historically the ADE Lie algebras were the first case to be made explicit (although other sets had of course been known for much longer). We will, as promised, return to introduce Lie theory in Section 3.5.

We have now discussed three different sets of root systems following ADE patterns. We will see later in Section 4.6 that there are three correspondences between pairs of these. We would like to stress here that the point of view of this book of the root system as being fundamental is somewhat unconventional. Traditionally, these three areas have been respectively considered as polytopes, subgroups of SU(2) and Lie algebras, and in those guises they seem very different things.

We advocate here that this different point of view can achieve a conceptual unification that might seem mysterious from a different perspective. In particular, much of Lie algebra theory – such as the root systems, Coxeter number and labels – are determined by more elementary concepts, which is lost in a lot of the Lie-centric literature. After all, since this book is aimed towards the beginning student, we think introducing root systems, which require only elementary linear algebra, is well justified. We will discuss some of these concepts now.

3.2.4 Coxeter Plane: Degrees and Exponents

We have seen above that Coxeter–Dynkin diagrams provide a simple way of visualising any root system. Most root systems can be somewhat unfamiliar, e.g., the root system of icosahedral symmetry is the icosidodecahedron and that of H_4 is the 600-cell. In particular, in higher dimensions these are hard to visualise.

However, somewhat surprisingly, there always exists a distinguished class of planes into which one can project, and which thus provides another canonical

---

to consider two infinite families and three exceptional cases in 3D – these in turn of course induce (determine) two infinite families and three exceptional cases in 4D: another ADE set. We will explore the connection between these, and the next (the actual ADE set), in Part II in Section 4.6.

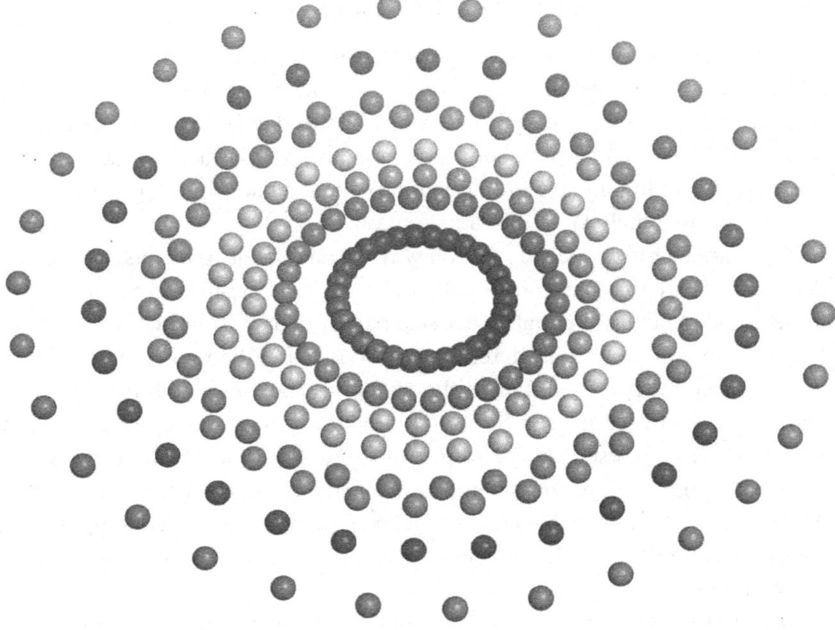

Figure 3.5 The projection of the 240 8D root vectors in the E_8 root system into the Coxeter plane is a popular visualisation of E_8 as eight concentric circles of 30.

way of visualising any root system as a 2D projection. For instance, Figure 3.5 shows a visualisation of E_8 [21] as the projection of the 240 root vectors in 8D into 8 concentric rings of 30 in the Coxeter plane. This plane has interesting properties connected with group invariants, which we will now explore. First, a definition

DEFINITION 3.7   The **Coxeter element** w of a Coxeter group is the product of all the simple reflections, i.e., $w = s_1 \ldots s_n$. It has the highest order of all the elements in the group, which is a useful invariant called the **Coxeter number** h.

Note that in the definition, the order in the product in w *does* matter, but all such Coxeter elements are conjugate to each other, and have the same Coxeter number and equivalent descriptions. Thus we will refer to w as *the* Coxeter element. Picking one such element, one can look for its eigenvectors and eigenvalues. Since it does not in general have real eigenvalues, people usually look for complex eigenvalues by complexifying the whole space.[14]

One can then show that there exists a distinguished plane in which this Coxeter element acts as an h-fold rotation [20]. Projection of a root system onto this Coxeter plane is thus a convenient way of visualising any finite Coxeter group in any dimension. In fact there are several such planes where the Coxeter element acts as an h-fold rotation, and one can look for eigenvalues of the form $\exp(2\pi i m_i/h)$. The Coxeter plane is the one distinguished by having $m = 1$ and $m = h - 1$. For instance, in the above E_8 example, the Coxeter number is 30 such that the Coxeter element rotates by a 30-fold rotation, i.e., an angle of $2\pi/30$, whilst its inverse rotates by 29 "notches" (or $2\pi \cdot 29/30$). However, as we shall see, there are also other planes in which they rotate by 7 and 23 notches, 11 and 19, or 13 and 17.

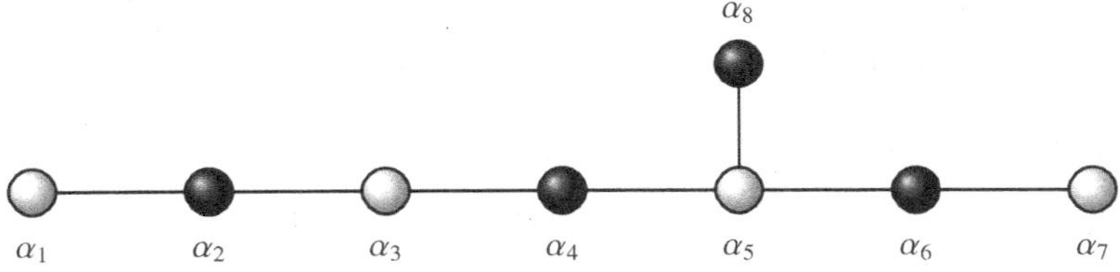

Figure 3.6 Illustration of the geometry of the Coxeter plane via bipartite colourings, diagram foldings and the Perron–Frobenius eigenvector.

The Coxeter element therefore factorises as the product of h-fold rotations in orthogonal planes. Clockwise and counterclockwise rotations in the same plane trivially yield exponents m and $h - m$ (by w and w^{-1}).

The usual construction of the Coxeter plane is via a two-fold colouring of the Dynkin diagram (see Figure 3.6). Since any finite Coxeter group has a tree-like diagram, one can partition the simple roots into two coloured sets (e.g., white and black) of roots, which are mutually orthogonal within each set. Since the Cartan matrix (for the spherical case here) is positive definite, there exists one eigenvector with all positive entries, which is called[15] the Perron–Frobenius eigenvector (see also Section 3.3).

This Perron–Frobenius eigenvector allows one to show the existence of the invariant Coxeter plane in the following way. One takes the reciprocals of the

---

[14] This is not strictly necessary, e.g., in Clifford algebra, as we discuss in Sections 4.4.1 and 4.4.2.

[15] It is expedient to remind the reader of the Perron–Frobenius Theorem: for an $n \times n$ positive matrix, there is an eigenvalue r whose absolute value is bigger than all others, whose corresponding eigenspace is 1-dimensional, and whose eigenvector has all positive entries.

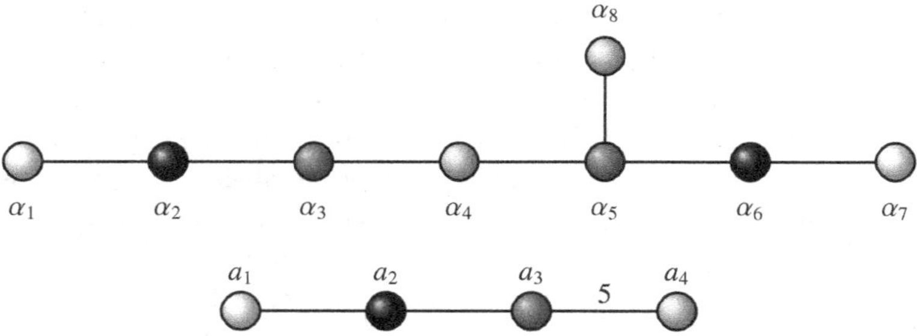

Figure 3.7 Coxeter–Dynkin diagram folding and projection from E_8 to H_4: one defines the new generators $s_{a_1} = s_{\alpha_1} s_{\alpha_7}$, $s_{a_2} = s_{\alpha_2} s_{\alpha_6}$, $s_{a_3} = s_{\alpha_3} s_{\alpha_5}$, $s_{a_4} = s_{\alpha_4} s_{\alpha_8}$, which themselves satisfy H_4 relations. At the level of roots, one essentially projects the eight simple roots of E_8 onto the four simple roots of H_4 and their τ-multiples.

simple roots (that is, a reciprocal basis ω_j such that $\alpha_i \cdot \omega_j \propto \delta_{ij}$ – these basis vectors are the weights) and then defines two distinguished vectors: a white vector that is a linear combination of the white weights with the corresponding coefficients taken from the Perron–Frobenius eigenvector, and a black one, which is a linear combination of the black weights with the right Perron–Frobenius coefficients. The Coxeter plane is then the plane defined by these two vectors,[16] in which the Coxeter element acts as an h-fold rotation, where h is the Coxeter number.

Just as an aside, the connection between (A_4, D_6, E_8) and (H_2, H_3, H_4) is related. It is in fact a 4-fold colouring, which could of course be easily collapsed to the 2-fold colouring giving the Coxeter plane. This is shown in Figure 3.7. The idea is that one defines 4 pairs of simple E_8 reflections that themselves satisfy H_4 relations [21], i.e., in a 4-dimensional subspace act as reflections.

The way the geometry of the Coxeter plane works is that the Coxeter element acts as an h-fold rotation in that plane, i.e., it rotates via angles $2\pi/h$ clockwise, as well as counterclockwise, which amount to exponents 1 and $h-1$. Other (conjugate) Coxeter elements will rotate by different amounts in that plane, or equivalently, there are other planes in which the Coxeter element still acts as an h-fold rotation, but with non-trivial exponents. These exponents are an additional set of invariants characteristic of each group, which we will encounter again in Section 4.4.2 and which we list in Table 3.2.

---

[16] In Clifford algebra, the outer product of two vectors defines a plane, so it will be the outer product of these two vectors that determines the Coxeter plane.

Type	h	m_i
A_n	$n+1$	$1, 2, \ldots, n$
B/C_n	$2n$	$1, 3, 5, \ldots, 2n-1$
D_n	$2(n-1)$	$1, 3, 5, \ldots, 2n-3, n-1$
E_6	12	$1, 4, 5, 7, 8, 11$
E_7	18	$1, 5, 7, 9, 11, 13, 17$
E_8	30	$1, 7, 11, 13, 17, 19, 23, 29$
F_4	12	$1, 5, 7, 11$
G_2	6	$1, 5$
H_2	5	$1, 4$
H_3	10	$1, 5, 9$
H_4	30	$1, 11, 19, 29$
$I_2(n)$	n	$1, n-1$

There is a close connection with other invariants of these groups, called degrees (of polynomial invariants) d_i, which are related to the above exponents m_i via

$$d_i = m_i + 1;$$

the proof involves the Coxeter plane [20]. Polynomial invariants are polynomials that are invariant under the group action. Their degrees are then also an invariant of the group. For instance, the symmetric group that exchanges x and y has invariant polynomials such as xy or $x^2 + y^2$, etc. The degrees of these invariants are fundamental properties of the symmetry groups. For instance, E_8 has invariant polynomials of degrees $2, 8, 12, 14, 18, 20, 24, 30$, as expected from the exponents we encountered above. We will mention invariants again in Section 5.8 in the context of complex polynomials. Suffice here to point out the close connection with exponents, as this will become important in Section 4.4.2. It is also interesting to mention that the product of the degrees is equal to the order of the Coxeter group $d_1 \ldots d_n = |W|$ and that $d_1 + \cdots + d_n = N + n$, where N is the number of (pure) reflections in the group, and n, the number of invariants, is equal to the rank of the root system. The interested reader can refer to [20] for further details.

This brief encounter with graphical representations of mathematical structures motivates the study of graphs in their own right, which we will now proceed to do.

3.3 Graphs and Spectra

The previous sections saw the emergence of ADE-ology from the perspective of polyhedra and root systems. We have seen that the previous mathematical structures such as root systems, Cartan matrices, etc., can be conveniently represented in diagrammatic form. We will now discuss graphs in their own right. As promised, we have kept the material at an undergraduate level to entice the young. It is indeed surprising that traditionally introductions of root systems and Coxeter groups are left to a graduate course on Lie theory, a tradition from which we purposefully broke (we defer the discussion till we have completed the more elementary material, that is, until Section 3.5).

Continuing in this vein, we now turn to some elementary graph theory, whence the ADE and affine ADE diagrams of Section 3.2 arise rather surprisingly and naturally. We see that they arise as the connected graphs with eigenvalues less than and equal to 2, respectively. From such a graph theory perspective, it should be absolutely clear that the diagrams are combinatorial objects, and that their defining property (underlying their occurrence throughout much of mathematics) is a simple fact of algebraic combinatorics.[17] Terry Gannon in [4] also considers these graphs as of fundamental importance, and calls them graphs with $PF2^-$ and $PF2$ assignments, respectively (Section 2.5.2 in [4]). We will see that in the context of the McKay correspondence in Section 4.5, these affine ADE/PF2 graphs arise very naturally precisely because of this property.

3.3.1 Largest Eigenvalue 2 or Smaller

Take a finite, connected, graph G. We recall that this is a collection of nodes with edges between some pairs of nodes. Here, we consider graphs which are *simple*, meaning that there is only at most a single edge between any pair of nodes, and *undirected*, meaning that we do not consider direction for the edges. We will consider directed graphs and associated quiver representations later in Section 5.4. For the graphs of our present concern, we first have that

DEFINITION 3.8   The **adjacency matrix** $\mathcal{A}$ of G is an $n \times n$ symmetric matrix, where n is the number of nodes and $\mathcal{A}_{ij} = 1$ if there is an edge between nodes i and j and 0 otherwise. The **spectrum** of G is that of $\mathcal{A}$ (the multiset of its eigenvalues).

Remark   In this section and throughout most of the book, we will only consider undirected graphs. In general, the edges in a graph could have direction

---

[17] For an introduction, by one of the authors, please see [31].

so that an arrow from node i to j corresponds to $A_{ij} = 1$. The adjacency matrix of a directed graph would not necessarily be symmetric.

The spectrum of a matrix, we recall, is the set of eigenvalues. The maximal eigenvalue is customarily called the *spectral radius*. Illustrative examples are given in Equation (3.9).

Furthermore, we make the distinction that if between two nodes, there are *multiple* edges (or arrows), then we would refer to the graph as a **multigraph**. In such cases, A_{ij} would equal the number of arrows from node i to node j. Indeed, since the subject of this book is ADE, the so-called simply-laced graphs, we won't be encountering multigraphs much at all. For a fairly self-contained rapid introduction to the terminologies for finite graphs, especially in the context of AI-assisted mathematics, the reader is referred to [32].

In summary, for us $\mathcal{A}$ is a binary symmetric square matrix; so its eigenvalues are real, which we can denote as

$$\lambda_1 \geq \lambda_2 \ldots \geq \lambda_n.$$

In particular, we will study λ_1 and λ_n, the largest and the smallest eigenvalues, in this book.

A remarkable theorem characterises how the largest eigenvalue is related to the (affine) ADE graphs:

THEOREM 3.9 (Smith [33]) *Let G be a finite connected graph with adjacency matrix $\mathcal{A}$, then*

- $\lambda_1 = \max\{\text{Eigenvalues}(G)\} < 2 \Leftrightarrow G$ *is ADE;*
- $\lambda_1 = \max\{\text{Eigenvalues}(G)\} = 2 \Leftrightarrow G$ *is Extended (affine) ADE.*

Whilst this theorem is implicit in the classification of Lie algebras, upon which we will expound shortly (see Section 3.5), the succinctness of its content is made ever more impressive by the fact that it should appear in its present form as late as 1969.

Let us give some illustrative examples to reify the above discussions. Consider the diagrams A_4 and $\widehat{A_3}$. These, and their respective 4×4 adjacency matrices are

$$A_4 \qquad \bullet\!\!-\!\!\!-\!\!\bullet\!\!-\!\!\!-\!\!\bullet\!\!-\!\!\!-\!\!\bullet \qquad \mathcal{A}_{A_4} = \begin{pmatrix} 0 & 1 & 0 & 0 \\ 1 & 0 & 1 & 0 \\ 0 & 1 & 0 & 1 \\ 0 & 0 & 1 & 0 \end{pmatrix}$$

$$\widehat{A_3} \qquad \mathcal{A}_{\widehat{A_3}} = \begin{pmatrix} 0 & 1 & 0 & 1 \\ 1 & 0 & 1 & 0 \\ 0 & 1 & 0 & 1 \\ 1 & 0 & 1 & 0 \end{pmatrix} \tag{3.9}$$

The two adjacency matrices differ only in the extremal principal off-diagonal entries and yet differ drastically in their eigenvalues:

$$\text{Eigenvalues}(\mathcal{A}_{A_4}) = \frac{1}{2}\left\{-1 - \sqrt{5}, 1 - \sqrt{5}, \sqrt{5} - 1, 1 + \sqrt{5}\right\} = \pm\{\tau, \sigma\}$$

$$\text{Eigenvalues}(\mathcal{A}_{\widetilde{A_3}}) = \{-2, 0, 0, 2\}. \tag{3.10}$$

Thus indeed the maximal eigenvalues are respectively < 2 and $= 2$, in agreement with Theorem 3.9.

Another way to state Theorem 3.9 is via *induced subgraphs*:

DEFINITION 3.10   An *induced subgraph* of a graph G consists of a subset of the vertex set together with all edges lying within this subset.

Indeed, the ADE graphs are induced subgraphs of the corresponding extended (affine) ADE graphs.

Thus, another way to state Smith's theorem is that for G connected,

THEOREM 3.11 $\lambda_1(G) \leq 2$ *if and only if* G *is an induced subgraph of an extended (affine) ADE graph.*

First, we note that

LEMMA 3.12   *For connected graphs* $G' \subsetneq G$, *the spectral radius* λ_1 *is strictly decreasing in the sense that* $\lambda_1(G') < \lambda_1(G)$.

This follows from the elementary fact that for non-negative matrices, the Perron–Frobenius theorem (which we recall from Theorems 2.18 and 2.19) guarantees the existence of $\lambda_1(G)$ with corresponding non-negative eigenvector x (respectively $\lambda_1(G')$ and corresponding x'). Thus

$$\lambda_1(G') = \frac{x'^T \mathcal{A}(G')x'}{x'^T x'} < \frac{x'^T \mathcal{A}(G)x'}{x'^T x'} < \max_{x \neq 0} \frac{x^T \mathcal{A}(G)x}{x^T x} = \lambda_1(G).$$

In particular, for an induced subgraph of G, its adjacency matrix is a principal submatrix of that of G, so its greatest eigenvalue does not exceed that of G.

Another useful result that follows from properties of non-negative matrices is that

LEMMA 3.13   *Let* G *be a connected graph and let* d_{min}, d_{max} *and* $\bar{d}$ *respectively be the minimal, maximal and average valency (number of edges adjacent to) of nodes, then*

$$d_{min} \leq \bar{d} \leq \lambda_1 \leq d_{max}.$$

We can formulate part of the Perron–Frobenius theorem which we require in terms of labellings, which we now define.

Let G be a connected graph with vertex set $\{1, 2, \ldots, n\}$, and λ a positive real number. A λ-*labelling* of G is an assignment of positive real numbers $x_1, x_2, \ldots, x_n$ to the vertices such that the sum of the labels on the neighbours of vertex i is λ times the label of i.

If $\mathcal{A}$ is the adjacency matrix of G, then the condition for a λ-labelling is simply the equation $Ax = \lambda x$, where x is the column vector with entries $x_1, \ldots, x_n$. Now the Perron–Frobenius theorem tells us the following:

- For any connected graph G, there is a unique λ for which G has a λ-labelling; moreover, λ is the spectral radius of G, and the λ-labelling is unique up to scalar multiple.
- The spectral radius of a proper connected induced subgraph of G is strictly smaller than that of G.

Here are the proofs.

PROPOSITION 3.14 *Let G be a connected graph which has a λ-labelling with respect to a and a μ-labelling with respect to b. Then*

- $\lambda = \mu$;
- *a and b differ by a constant factor.*

Proof We use the notation $v \sim w$ to mean that v and w are joined by an edge (adjacent) in the graph. Now we have

$$\sum_{w \sim v} a(w) = \lambda a(v) \tag{3.11}$$

for fixed v, and similarly

$$\sum_{v \sim w} b(v) = \mu b(w) \tag{3.12}$$

for fixed w. So

$$\lambda \sum_{v \in V(\Gamma)} a(v)b(v) = \sum_{w \sim v} a(w)b(v) = \mu \sum_{w \in V(\Gamma)} a(w)b(w)$$

where the middle sum is over all adjacent pairs (v, w). Since the values of a and b are positive, this forces $\lambda = \mu$.

For the second part, observe that $c = a - xb$ is also a λ-labelling for any real number x, as long as $c(v) > 0$ for all v. Let s be the supremum of all such values of x. Then $c(v) \geq 0$ for all v, but there exists w such that $c(w) = 0$. Now the sum of $c(v)$ over the neighbours v of w is zero, so all such values $c(v)$ are zero. Using the connectedness of the graph, it now follows that $c(v) = 0$ for all vertices v, so that $a = sb$, as required. $\qquad\square$

PROPOSITION 3.15   *Suppose that $\mathcal{G}'$ and $\mathcal{G}$ are connected graphs, with $\mathcal{G}'$ a proper induced subgraph of $\mathcal{G}$. Then $\mathcal{G}'$ and $\mathcal{G}$ cannot both have a λ-labelling.*

Proof   Suppose for a contradiction that a is a λ-labelling of $\mathcal{G}'$, and b a λ-labelling of $\mathcal{G}$. In what follows, *summations are over vertices of $\mathcal{G}'$ only*. Equation (3.11) still holds; but in place of Equation (3.12) we have strict inequality:

$$\sum_{v \sim w} b(v) > \lambda b(w) \tag{3.13}$$

for fixed w. This is because we sum only over the neighbours of w in $\mathcal{G}'$, and at least one vertex of $\mathcal{G}'$ has a neighbour outside $\mathcal{G}'$, since $\mathcal{G}'$ is connected. So we have

$$\lambda \sum_{v \in V(\mathcal{G}')} a(v)b(v) = \sum_{w \sim v} a(w)b(v) > \lambda \sum_{w \in V(\mathcal{G}')} a(w)b(w),$$

a contradiction. $\square$

With this in hand, we give the proof of Smith's theorem. We have to show the following:

- The spectral radius of an ADE graph is less than 2, and any connected graph with spectral radius less than 2 is an ADE graph.
- The spectral radius of an affine ADE graph is 2, and any connected graph with spectral radius 2 is an affine ADE graph.

We show that the spectral radius of an affine ADE graph is 2 by giving a 2-labelling of each of these graphs. For $\widehat{A}_n$ (an $(n + 1)$-cycle), we can label each vertex by 1. The other cases are given in Table 3.2. For convenience, we give the labelling of $\widehat{E}_8$ here.

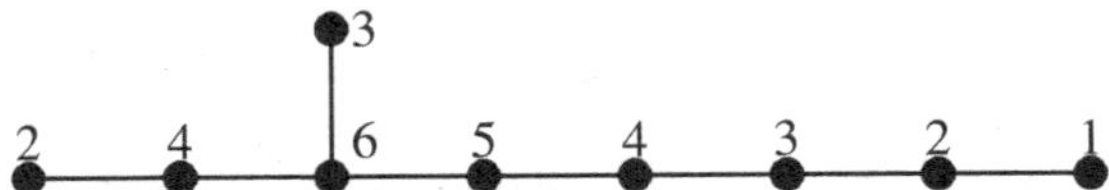

It follows from Proposition 3.15 that the spectral radius of an ADE graph is strictly less than 2.

Now we give the proof of the converse. Suppose that $\mathcal{G}$ has spectral radius strictly less than 2. Then $\mathcal{G}$ cannot contain any affine ADE graph. In particular, it does not contain $\widehat{A}_n$ (a cycle), for any n; so it is a tree. It does not contain $\widehat{D}_n$, and so cannot have more than one branchpoint, or a branchpoint lying on more than three edges. If it has no branchpoint, it is a path A_n; otherwise, it doesn't contain $\widehat{E}_n$, so the lengths of the three arms are restricted to $(2, 2, n)$, $(2, 3, 3)$, $(2, 3, 4)$ or $(2, 3, 5)$, giving us the D_n and E_n diagrams.

Finally we turn to the proof that the affine ADE graphs are the only graphs with spectral radius 2. Let G be a graph with spectral radius 2. Then the subgraph obtained by deleting any vertex has spectral radius strictly less than 2, and therefore is an ADE graph.

If G has a cycle, then it contains (and so is equal to) $\widehat{A_n}$. Otherwise, it is a tree. If it contains two different branchpoints, or a branchpoint with valency greater than 3, then it contains (and so is equal to) $\widehat{D_n}$ for some n. In the remaining cases, the lengths of the three arms are easily found, and the graph is $\widehat{E_n}$ for some n.

Remark Hidden here is a remarkable property of spectral radius 2. Any graph with spectral radius less than 2 can be embedded as an induced subgraph in a graph with spectral radius 2.

Remark Part 1 of Theorem 3.9 implies that the ordinary ADE diagrams are the only ones which admit labels so that twice the label of each node minus 2 is the sum of adjacent labels. Note that Gannon calls these $PF2^-$ assignments, reflecting that the eigenvalue less than 2 property and this modified labelling are also closely related [4]). Note that Gannon suggested that all ADE patterns might fundamentally reduce to these two ADE/affine ADE or $PF2^-/PF2$ multigraph classifications (but notes the notable potential exception of modular invariants in Section 5.8.) These labels also appear in the half-sum of positive roots as twice the coefficients in the simple roots [34] though it doesn't seem obvious what the significance of this is.

The fact that the ADE diagrams have greatest eigenvalue less than 2 while the extended diagrams have greatest eigenvalue 2 means that the difference $2I - \mathcal{A}$ between twice the identity matrix and the (symmetric) adjacency matrix is positive definite for ADE and positive semi-definite for extended ADE. This difference is none other than the Cartan matrix discussed in Section 3.2.1.

Now, a positive semi-definite real symmetric matrix is the *Gram matrix* of a set $\{v_1, \ldots, v_n\}$ of vectors in a real inner product space, that is, its (i, j) entry is the scalar product $v_i^T \cdot v_j$:

$$(2I - \mathcal{A})_{ij} = v_i^T \cdot v_j \, .$$

In our case, since $\mathcal{A}$ has zero diagonal (there are no self-adjoining loops on any node) and all entries are 0 and 1 (we do not have multigraphs), the diagonal entries of $2I - \mathcal{A}$ are all equal to 2, while the off-diagonal entries are 0 or -1. This means that each vector v_i has length $\sqrt{2}$, and any two vectors lie at an angle of $\cos^{-1}(0) = 90°$ or $\cos^{-1}(-1 \cdot (\frac{1}{\sqrt{2}})^2) = 120°$. We saw the significance of this when we encountered root systems, where we preferred the unit normalisation for simplicity with reflections. The above normalisation is customary in Lie theory, however.

Having expounded upon λ_1, what about the "dual" problem – connected graphs whose adjacency matrix has least eigenvalue -2 or greater? In our notation above, we need to consider graphs with $\lambda_n \geq -2$. Whilst the statement of this dual problem is intuitively easy and analogous to the above, this was a topic of great interest in the 1960s with a more subtle answer [35]. We defer a treatment till the more advanced Section 5.3.

3.3.2 Graph Laplacians

One can rephrase the above discussion in yet another way. First, the degree matrix $\mathcal{D}$ of a graph $\mathcal{G}$ is the diagonal matrix recording the total degree of each node (note that a loop on a node, should it exist, contributes 2). Then, we have the definition (q.v. [32])

DEFINITION 3.16 The (graph or discrete) **Laplacian** Δ for a graph is the difference between the degree matrix $\mathcal{D}$ and the adjacency matrix $\mathcal{A}$:

$$\Delta := \mathcal{D} - \mathcal{A} . \tag{3.14}$$

Therefore, for a function $\phi \colon V \to R$ taking the vertices V to an appropriate ring R, $[\Delta(\phi)](v) = \sum_w (\phi(v) - \phi(w))$ for all vertices w adjacent to $v \in V$. Theorem 3.9 can then be recast into the statement that the extended ADE diagrams are the only ones for which the eigenvalue-one problem for the discrete Laplacian can be solved:

$$\Delta\phi = \phi . \tag{3.15}$$

3.4 Binary Polyhedral Groups and SU(2)

Now, since we are armed with some knowledge of both finite and continuous groups from Chapter 2, we can move on to the subject of (discrete) finite subgroups of these continuous groups.[18] This is an interesting and wide topic, but once again, we will be interested in the classification thereof and the ADE pattern which emerges. Furthermore, we will see how Clifford algebras provide a simpler and more unified way of viewing some of these standard topics.

3.4.1 The Double Cover of SO(3): Spin(3), SU(2), and Quaternions

Let us return to our familiar group SO(3), the rotations of $\mathbb{R}^3$ which set the scene for the Platonic solids. We now know that it is a continuous group, and also a Lie group. It is connected as a manifold, but not simply connected (as even O(3) is already not connected). The simply connected universal cover of SO(3) is a manifold called Spin(3). It is a double cover and it is in fact isomorphic to SU(2). We emphasise that these statements are on the level of the *group*. At the level of the Lie algebra, $\mathfrak{su}(2) = \mathfrak{so}(3)$, as one can see from the Dynkin diagrams: they share the Dynkin diagram of type A_1, a single node. To see the double cover at the group level, we first recall the quaternions.

The quaternions are a generalisation of the complex numbers, which are generated by the unit imaginary i with $i^2 = -1$ (though we will have something to say about this in later Clifford algebra Sections 3.4.4 and 4.4.2). In the quaternions, there are three imaginary units $i^2 = j^2 = k^2 = -1$, and their interplay is regulated by $ij = k = -ji$ (and cyclic permutations thereof).

Willam Rowan Hamilton spent some time trying to construct a system with two imaginary units. While walking by a canal in Dublin, he had the inspiration to use three, and scrawled the equations on Brougham Bridge, which now has a commemorative plaque to mark the occasion (Figure 3.8).

More formally,

DEFINITION 3.17 (Quaternions) The quaternions $\mathbb{H}$ are a 4-dimensional algebra over $\mathbb{R}$ defined by

$$\mathbb{H} = q = a + bi + cj + dk \ : \ a, b, c, d \in \mathbb{R}$$

---

[18] These are also Lie groups, i.e., also have a manifold structure. A lot of the concepts can be understood by simply considering the continuous group structure without an in-depth understanding of the manifold structure, which will be further discussed in the next part on Lie theory, Section 3.5.

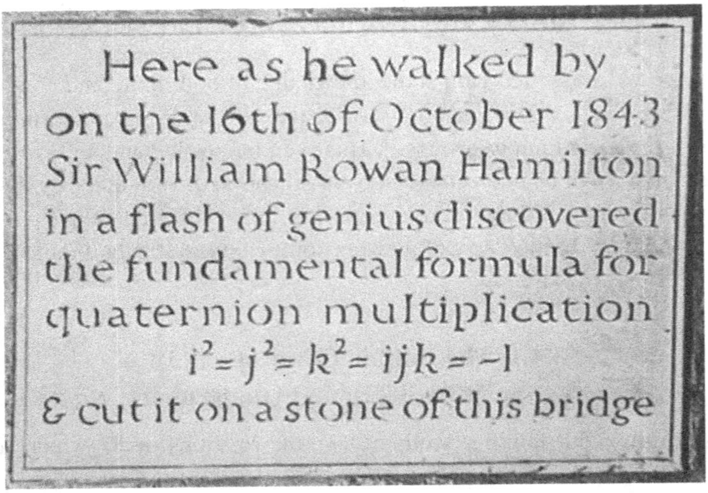

Figure 3.8 Hamilton's discovery of the quaternions. Image from MacTutor
`https://mathshistory.st-andrews.ac.uk/Biographies/Hamilton/`.

with

$$i^2 = j^2 = k^2 = -1; \ ij = -ji = k, \ ki = -ik = j, \ jk = -kj = i \,.$$

In analogy with the complex numbers, we can define a quaternionic conjugate $\bar{q} = a - bi - cj - dk$ and a norm by $|q|^2 = q\bar{q} = a^2 + b^2 + c^2 + d^2$. This turns $\mathbb{H}$ into a normed division algebra, and means that we can think of the quaternions in some sense as 4D space, $\mathbb{R}^4$, or even 4D Euclidean space. The unit sphere in $\mathbb{H}$ $a^2 + b^2 + c^2 + d^2 = 1$ is clearly the 3-sphere S^3 which is also the same as SU(2) as a manifold. Specifically,[19] we have

THEOREM 3.18   SU(2) *is a double cover of* SO(3) *as Lie groups.*

Proof   One takes an element of SU(2), for $a, b, c, d \in \mathbb{R}$ and can map it into $\mathbb{H}$ as

$$\text{SU(2)} \ni M = \begin{pmatrix} a + bi & c + di \\ -c + di & a - bi \end{pmatrix} \longrightarrow a + bi + cj + dk \in \mathbb{H} \,. \tag{3.17}$$

---

[19] The fancier way of saying this is that SO(3) admits the central extension $\mathbb{Z}/2\mathbb{Z}$ to SU(2) via the short exact sequence

$$0 \longrightarrow \mathbb{Z}/2\mathbb{Z} \longrightarrow \text{SU(2)} \longrightarrow \text{SO(3)} \longrightarrow 0 \,. \tag{3.16}$$

We can check that the SU(2) conditions $\overline{M}^T M = \mathbb{I}_2$ and $\det(M) = 1$ simply translate to $a^2 + b^2 + c^2 + d^2 = 1$, which is a 3-sphere. In other words, as mentioned earlier, SU(2) $\simeq S^3$ topologically.

There is thus a two-to-one homomorphism from the special unitary group SU(2) to the rotation group SO(3). $\qquad\qquad\square$

Spin Groups: In general, SO(n) is doubly covered by Spin(n) (more on this in the next section on Clifford algebras), whilst O(n) is doubly covered by Pin(n)[20]; the isomorphism of Spin(3) to SU(2) is accidental. Spin(3) is also accidentally isomorphic to the quaternions or Sp(1), so all these are pretty interchangeable in 3D. However, as we will see in the next section, in general it is geometrically clearer to think in terms of Clifford algebras and Spin groups. We now discuss the generic case of the spin groups, which are easily constructed in Clifford algebra, as these are the concepts that generalise straightforwardly, as opposed to accidental isomorphisms. But first, let us finish off our discussion on the Platonic solids.

3.4.2 Discrete Finite Subgroups of SU(2)

We learnt that the regular discrete symmetries of SO(3) are the ADE polyhedral groups as finite rotation groups G. Due to Theorem 3.18, these lift to their double covers $\tilde{G}$ in SU(2), i.e., a group with a centre Z of order 2 such that $\tilde{G}/Z = G$, Note, incidentally, that Z contains the unique involution in $\tilde{G}$ (since $-I$ is the only involution in SU(2)). These double covers of the rotational polyhedral groups are called the **binary polyhedral groups**, and have orders twice that of the rotational groups. They have the same order as the corresponding Coxeter groups (the full polyhedral groups) but are of course different groups, as the Coxeter goups live in O(3) and contain the polyhedral groups with index 2 while the binary polyhedral goups are double covers of then in Spin(3) = SU(2)).

The modern parlance of stating the classification problem in (3.1) is that we are identifying the discrete finite subgroups of the continuous – and Lie – group SO(3), the isometry group of $\mathbb{R}^3$. Theorem 3.18 (and the short exact sequence (3.16)) implies that all the groups in (3.1) lift to finite subgroups of SU(2). In the abelian case of the cyclic group $\mathbb{Z}/n\mathbb{Z}$, the lift is trivial, in all other cases, the lift is non-trivial. We will encounter a simple construction of this binary

---

[20] Note that this terminology is slightly tongue in cheek: since often the condition of determinant being equal to 1 is denoted by the label S (e.g., in SL(3)), and the Spin group denoting rotations (i.e., det 1), calling its double cover Pin encapsulates the idea of relaxing to det ± 1 in a humorous way.

double cover in Section 3.4.4 on Clifford algebras. Again, we emphasise that this is a double cover, with the same order as but different from the groups obtained by including reflections, which are subgroups of O(3).

We shall make important use of these binary groups when we discuss ADE correspondences, and in particular the McKay correspondence, and thus summarise them here:

	Presentation of G	$\|G\|$
Cyclic	$\langle R \mid R^n = \mathbb{I} \rangle \simeq \mathbb{Z}/n\mathbb{Z}$	n
Binary dihedral	$\langle R, S \mid R^n = S^2, \; S^{-1}RSR = R^{2n} = \mathbb{I} \rangle$	$4n$
Binary tetrahedral	$\langle R, S, T \mid RST = R^2 = S^3 = T^3 \rangle$	24
Binary octahedral	$\langle R, S, T \mid RST = R^2 = S^3 = T^4 \rangle$	48
Binary icosahedral	$\langle R, S, T \mid RST = R^2 = S^3 = T^5 \rangle$	120

$$(3.18)$$

We see that all the non-abelian cases have the order doubled and the presentations differ from (3.1) merely by dropping the equality to the identity ($RST = \mathbb{I}$ from Table 3.1). In fact, since there is a unique involution, this presentation can also be taken as $RST = -\mathbb{I}$. In particular, while $R^n = \mathbb{I}$ for the cyclic case, this is not so for the others. Furthermore, in the binary tetrahedral, octahedral and icosahedral cases, the generator R is redundant – it being expressible in terms of S and T – but we keep it for reference.

These binary polyhedral groups have various presentations. For example, to many, they are known as discrete subgroups of SU(2), so just a finite set of 2×2 complex matrices. When seen as discrete groups of unit quaternions, the binary groups have the names of Hurwitz units, Lipschitz units and the Icosians. We will also see another incarnation of the binary polyhedral groups as multivector groups in Clifford algebras shortly.

3.4.3 A Bit of Polyhedral Group Theory

Since we will encounter these binary polyhedral groups again in greater detail later, we quickly summarise their main properties, as well as recalling some standard results (see also Chapter 2).

First, we recall some standard (q.v. [14]) facts from representation theory of a finite group G:

- The number n of irreducible representations (irreps) is given by the number of conjugacy classes:

$$n = \#\{\mathrm{Conj}(G)\} = \#\{\mathrm{Irrep}(G)\}.$$

- Clearly, $|G| = \sum_{i=1}^{n} |\mathrm{Conj}_i(G)|$, the size of the group is the sum over the sizes of the conjugacy classes. Moreover, each $|\mathrm{Conj}_i(G)|$ divides $|G|$.

- We let d_i denote the dimension of $\text{Irrep}_i(G)$, then d_i also divides $|G|$ and the sum of the squares of the dimensions of these irreducible representations add up to the order of the group

$$\sum_{i=1}^{n} d_i^2 = |G|.$$

- Of course, the trivial representation is always one of $\text{Irrep}_i(G)$, so at least one of the d_i is 1.

EXAMPLE 3.19 (Icosahedral and binary icosahedral groups)   For instance, the icosahedral group $I = A_5$ has five conjugacy classes: the identity, 15 2-fold rotations, 20 3-fold rotations, and two sets of 12 5-fold rotations. It is of order 60, which implies that it has five irreducible representations of dimensions 1, 3, $\bar{3}$, 4 and 5. Its double cover, the binary icosahedral group $2I$, is of order 120 and has an additional 4 conjugacy classes, i.e., 9 in total. It therefore has a further four irreducible spinorial representations 2_s, $2'_s$, 4_s and 6_s (with $2^2 + 2^2 + 4^2 + 6^2 = 60$ as required). This set of integers $1, 2, 3, 4, 5, 6, 4, 3, 2$ might already look familiar, e.g., from Section 3.2.2.

This binary icosahedral group has a mysterious connection with E_8 via the so-called McKay correspondence [3]. This actually applies to all other binary polyhedral groups and the root systems of ADE-type. This connection will be discussed in Section 4.5, and as we hinted at previously is usually cast in the language of (affine) Lie algebras. However, most of the structure is already contained at the level of the root systems.

EXAMPLE 3.20 (Octahedral groups)   The octahedral group O of order 24 consists of the identity, six 2-fold rotations, eight 3-fold rotations, six 4-fold rotations and another class of three 2-fold rotations around those 4-fold axes. The dimensions of the irreps are therefore $1, 1, 2, 3, 3$. The doubled binary octahedral group has another 3 conjugacy classes, giving further irreps 2_s, $2'_s$ and 4_s. As expected, these are the same labels we have already encountered for E_7.

EXAMPLE 3.21 (Tetrahedral groups)   The four conjugacy classes of T of order 12 (two quartets of 3-fold rotations and three 2-fold rotations) determine that this group has four irreducible representations of dimensions $1, 1', 1''$ and 3. The seven conjugacy classes of $2T$ of order 24 mean that this acquires a further three irreducible spinorial representations of dimensions 2_s, $2'_s$, $2_s''$. The dimensions of the irreps of the binary tetrahedral group are thus reminiscent of the E_6 labels.

The D_n family corresponds to the binary dihedral groups (aka the dicyclic groups), whilst the binary preimage of a cyclic group is a cyclic group of twice the order. Since cyclic groups are abelian, each element is in a conjugacy class on its own, and thus all the dimensions of the irreps and corresponding diagram labels are 1, with the irreps given by the nth roots of unity.

This concludes our discussion of discrete subgroups of SU(2), the binary polyhedral groups, with a view to making the connection with ADE via the McKay correspondence in Section 4.5.

3.4.4 Clifford Algebras

Root systems are a useful paradigm for reflection groups. As hinted at the outset, Clifford algebras [36–40] are also very efficient at performing reflections and are in fact very natural – perhaps the most natural – objects to consider in this framework [16]: the definition of a root system only stipulated a vector space with an inner product (symmetric bilinear form). That is exactly the structure one needs to define a Clifford algebra over this vector space. So without loss of generality we can construct this Clifford algebra over the vector space by using the inner product.

We therefore proceed to define an algebra as follows:

DEFINITION 3.22　　(Clifford Algebra) We have a vector space of elements $x, y, \ldots$ with

Product $xy = x \cdot y + x \wedge y$ (the algebra product, or "geometric product");

Inner Product This is the symmetric part of the algebra product, which is given by the scalar product on the vector space: $x \cdot y = (x|y) = \frac{1}{2}(xy + yx)$ and produces scalar, which is given by the symmetric bilinear form on the original vector space;

Exterior Product This is the antisymmetric part $x \wedge y = \frac{1}{2}(xy - yx)$ of the algebra product and is also called an outer product, with the resulting product being called a bivector. This is also familiar from the exterior algebra.

Remark: The exterior product denotes the oriented area spanned by the two constituent vectors, which gets reversed under interchange of x and y and denotes the plane spanned by them. These symmetry properties mean that parallel vectors commute whilst orthogonal vectors anticommute. The exterior product is of course in three dimensions also related to the perhaps more familiar vector cross product, since each plane (bivector) has a unique normal direction.

EXAMPLE 3.23 (Parallel vectors commute)   We now show that parallel vectors commute. Vectors being parallel means that $y = \alpha x$ for some scalar α. However, $x \wedge x = -x \wedge x = 0$, because the exterior product is antisymmetric. So we have

$$xy = x \cdot y + x \wedge y = x \cdot y + \alpha x \wedge x = x \cdot y = y \cdot x = y \cdot x + y \wedge x = yx.$$

Thus parallel vectors commute simply because the scalar product is symmetric and the two factors in it commute in the first place.

EXAMPLE 3.24 (Perpendicular vectors anticommute)   Conversely, we now show that perpendicular vectors anticommute. Vectors being perpendicular means that $x \cdot y = 0$. So we have

$$xy = x \cdot y + x \wedge y = 0 + x \wedge y = -y \wedge x = -y \wedge x - y \cdot x = -yx.$$

Thus perpendicular vectors anticommute simply because of the antisymmetry of the exterior product.

We extend the algebra product via linearity and associativity. This enlarges the algebra to a 2^n-dimensional vector space, which is isomorphic to the familiar exterior algebra as a vector space, though they are not isomorphic as algebras. In fact, the Clifford algebra is much richer, since the algebra product is invertible (at least for non-null vectors).

EXAMPLE 3.25 (Clifford algebra of 2D)   The Clifford algebra of two orthogonal unit vectors e_1 and e_2 (i.e., $e_1^2 = e_2^2 = 1$ for 2D Euclidean space) is 4-dimensional,

$$\underbrace{\{1\}}_{1 \text{ scalar}} \quad \underbrace{\{e_1, e_2\}}_{2 \text{ vectors}} \quad \underbrace{\{e_1 e_2\}}_{1 \text{ bivector}} .$$

The only new object here is $e_1 e_2$ (often denoted by I in this context).[21] This bivector defines the plane given by e_1 and e_2. It turns out to square to -1 and describe rotations in this 2D plane:

$$(e_1 e_2)^2 = (e_1 e_2)(e_1 e_2) = e_1(e_2 e_1)e_2$$
$$= e_1(-e_1 e_2)e_2 = -(e_1)^2(e_2)^2 = -1^2 \cdot 1^2 = -1.$$

EXAMPLE 3.26 (Rotations in 2D)   If we have a vector $x = x_1 e_1 + x_2 e_2$ in this plane then we have

$$Ix = e_1 e_2(x_1 e_1 + x_2 e_2) = x_1 e_1 e_2 e_1 + x_2 e_1 e_2 e_2$$
$$= x_1 e_1(-e_1 e_2) + x_2 e_1(e_2 e_2)^2 = -x_1 e_2 + x_2 e_1.$$

---

[21] Since $e_1^2 = e_2^2 = 1$, multiplying with more e_1s or e_2s doesn't actually lead to anything else.

This is actually a rotation by $\frac{\pi}{2}$ counterclockwise! If we think about xI, we see that one of e_1 and e_2 in e_1e_2 will commute with the vector part whilst the other anticommutes, so that overall $Ix = -xI$ also anticommute.

This 2D case is actually somewhat misleading, as we will see below.

e_1e_2 can thus play the roles of the usual imaginary unit i in complex numbers. But it has more geometric meaning. We will see more of these occurrences later where different unit imaginaries describe different planes. The spaces of vectors and that of the spinors (the even subalgebra) are both of dimension two, and can be mapped into each other, e.g., by multiplication by e_1: if $R = a_1 + a_2e_1e_2$ then $Re_1 = a_1e_1 - a_2e_2$.

One can define the exponential of an element in this algebra in the usual way via a series expansion. The finite 2^n-dimensionality means that only 2^n algebraically different terms contribute, whilst the coefficients add up as series expansions: one ends up with a spherical, Euclidean or hyperbolic trigonometric expansion, depending on whether the object one exponentiates squares to something positive, null or negative. We will encounter examples of bivector exponentials below in the context of the Coxeter plane.

EXAMPLE 3.27 (Spinors in 2D)   A spinor $a_1 + a_2e_1e_2$ can be written as a bivector exponential $\exp(\phi e_1e_2) = \cos\phi + e_1e_2\sin\phi$, via the usual Euler formula. One here sees the usual vector and spinor-like aspects of complex numbers disentangled, but related in a coherent framework. Transformation of a vector x by double-sided action of a bivector exponential leads to

$$\exp(\phi e_1e_2)x\exp(-\phi e_1e_2)) = \exp(2\phi e_1e_2)x,$$

since in 2D e_1e_2 anticommutes with a vector. That is, double-sided action of a spinor with angle ϕ leads to a rotation of a vector[22] by angle 2ϕ (it is customary to halve these angles and speak of a half-angle formula). This double-sided action is the concept that generalises. It is only accidentally the same as left-multiplication by a complex number in 2D, since in general the pseudoscalar/rotation plane bivector e_1e_2 does not necessarily (anti)commute with every vector.

Since a spinor $\exp(\phi e_1e_2)$ leads to a rotation by twice the angle, if we take $\phi = \pi$, then the overall rotation is one by 2π. A vector x is thus returned to its original state. However, the spinor itself is not!

$$\exp(\pi e_1e_2) = \cos\pi + \sin\pi e_1e_2 = -1 = \exp(-\pi e_1e_2).$$

---

[22] Of course, spinors multiply single-sidedly, and form a group under such multiplication.

It would require a rotation through 4π to return the spinor to its original state. That is exactly why these objects are named spinors. Spinors are often thought to be inherently quantum mechanical. But here we see them as a simple question of geometry. Composition of rotations (given by double-sided action) is achieved by composition of spinors by the geometric product.

Let us look at this curious double-sided transformation in more generality:

A General Treatment of Reflections

Using the above form $x \cdot y = (x|y) = \frac{1}{2}(xy + yx)$ for the inner product in the (simple) reflection formula Equation (3.5) and assuming unit normalisation of roots α_i one gets the much simplified version for the reflection formula

$$s_i: x \to s_i(x) = -\alpha_i x \alpha_i$$

since

$$x' = x - 2(x \cdot n)n = x - 2\frac{1}{2}(xn + nx)n = x - x - nxn = -nxn = -(-n)x(-n)$$

since $n^2 = 1$. This is much simpler than the usual reflection formula in Section 3.2.1. The only alternative system that comes close to this efficiency are the quaternions (see previous section), where it is seen as miraculous. However, for some historic reason the same sense of wonder is not usually extended to the Clifford version of this formula which holds in any dimension and signature, and is only accidentally isomorphic to the quaternions in 3D.

EXAMPLE 3.28 (A reflection) Let us consider such a reflection of a 2D vector $x = x_1 e_1 + x_2 e_2$ in the hyperplane (line) defined by e_1:

$$x' = -e_1 x e_1 = -e_1(x_1 e_1 + x_2 e_2)e_1 = -e_1 e_1(x_1 e_1 - x_2 e_2) = -x_1 e_1 + x_2 e_2.$$

That is, the component along e_1 gets reversed, whilst other directions are not affected. It is a simple exercise to translate this into matrix language.

Moreover, since via the Cartan–Dieudonné theorem most "interesting" (at least from a mathematical physics perspective: orthogonal, conformal, modular) groups can be written as products of reflections [16, 40], this formula actually provides a completely general way of performing such transformations by successive multiplication with the unit vectors defining the reflection hyperplanes

$$s_1 s_2 \ldots s_k: x \to s_1 s_2 \ldots s_k(x) = (-1)^k \alpha_1 \alpha_2 \ldots \alpha_k x \alpha_k \ldots \alpha_2 \alpha_1 =: \pm A x \tilde{A}.$$

A is defined as the product of the vectors which define the hyperplanes, so $A = \alpha_1 \alpha_2 \ldots \alpha_k$, and the tilde denotes the reversal of the order of the constituent vectors in the product (an involution). In order to study the groups of transformations, one therefore only needs to consider products of (root) vectors in the Clifford algebra. This is therefore an extremely (if not completely) general way of doing group theory.

EXAMPLE 3.29 (A rotation) Let us now consider a rotation such as a double-reflection of a 2D vector $x = x_1 e_1 + x_2 e_2$ in the hyperplanes (lines) defined by first e_1 and then e_2:

$$x' = -e_1 x e_1 = -e_1(x_1 e_1 + x_2 e_2)e_1 = -e_1 e_1(x_1 e_1 - x_2 e_2) = -x_1 e_1 + x_2 e_2.$$

$$x'' = -e_2 x' e_2 = -x_1 e_1 - x_2 e_2 = -x.$$

That is, both the components now get reversed. This therefore amounts to a rotation in the $e_1 e_2$-plane by π. It is a simple exercise to translate this into matrix language. Care, that in higher dimensions, only the 1- and 2-directions would be reversed, still giving a rotation in the $e_1 e_2$-plane by π, whilst other dimensions would not be affected (since they anticommute with both e_1 and e_2).

Since n and $-n$ encode the same reflection (as we noted above), products of unit vectors are double covers of the respective orthogonal transformation, as A and $-A$ encode the same transformation. We call even products R, i.e., products of an even number of vectors, spinors or rotors, and a general product A versors or pinors. They form the Pin group and constitute a double cover of the orthogonal group O(n), whilst the even products form the double cover of the special orthogonal group SO(n), called the Spin group. Clifford algebra therefore provides a particularly natural and simple construction of the Spin groups.

In fact, the above extends to an even more general theorem on the Clifford Algebra representation of orthogonal transformations. A versor is a multivector $A = a_1 a_2 \ldots a_k$ which is the product of k non-null vectors a_i ($a_i^2 \neq 0$). These versors also form a multiplicative group under the geometric product, called the versor group, where inverses are given by $\tilde{A}$, scaled by the magnitude $|A|^2 := |a_1|^2 |a_2|^2 \ldots |a_k|^2 = \pm A\tilde{A}$, where the sign depends on the signature of the space. The inverse is simply $A^{-1} = \pm \frac{\tilde{A}}{|A|^2}$. The Versor Theorem [41, 42] then states that

THEOREM 3.30 (versor theorem) *Every orthogonal transformation $\underline{A}$ of a vector v can be expressed via unit versors in the canonical form*

$$\underline{A}: v \to v' = \underline{A}(v) = \pm \tilde{A} v A, \tag{3.19}$$

where the $\pm$-sign defines its parity.

Unit versors are thus double-valued representations of the respective orthogonal transformation.

In particular, we get the binary polyhedral groups, the double cover of the rotational polyhedral groups, merely by multiplying together root vectors in the Clifford algebra (see Section 4.4.1 for more details). Thus the remarkably simple construction of the binary polyhedral groups (which are the spin double covers of the polyhedral groups) in our context is not at all surprising from a Clifford point of view.

Scalars, Bivectors, Trivectors, ...

Whilst the above discussion was completely general for orthogonal groups in spaces of arbitrary dimension and signature (and via some isomorphisms also the conformal and modular groups [16, 43, 44]), here we are particularly interested in 3D reflection groups. We will therefore only further consider 3D and 4D geometry here, extending our discussion of the 2D case above.

EXAMPLE 3.31   The Clifford algebra of 3D is generated, e.g., by three orthogonal unit vectors e_1, e_2 and e_3. This yields an eight-dimensional vector space consisting of the elements

$$\underbrace{\{1\}}_{\text{1 scalar}} \quad \underbrace{\{e_1, e_2, e_3\}}_{\text{3 vectors}} \quad \underbrace{\{e_1 e_2 = I e_3, e_2 e_3 = I e_1, e_3 e_1 = I e_2\}}_{\text{3 bivectors}} \quad \underbrace{\{I \equiv e_1 e_2 e_3\}}_{\text{1 trivector}}.$$

Any of the bivectors or trivectors square to -1. Thus, one now gets several different imaginary units based on a real vector space – without complexifying the whole space. In fact, one needs to be careful with such imaginary units since they do not necessarily (anti)commute. In fact, the scalar and the three bivectors satisfy quaternionic relations (see preceding section). Often the appearance of quaternions in mathematical physics just signals that the spin group Spin(3) is involved. Similarly, these commutation relations of 3 orthogonal unit vectors e_1, e_2 and e_3 have a matrix representation that many readers will be familiar with: the Pauli matrices. Therefore, the appearance of Pauli matrices often just heralds the same involvement of the spin group Spin(3). However, it is not inherently quantum mechanical, but purely arises from the geometry of a vector space with a scalar product.

Thus, the other two normed division algebras $\mathbb{C}$ and $\mathbb{H}$ thus emerge naturally within real Clifford algebras, without the need to complexify or quaternionify the whole underlying vector space. We see in the context of the Coxeter plane that this is actually much more natural and geometrically insightful (see Sections 3.2.4 and 4.4.2). The geometric interpretation of these elements of the Clifford algebra is still as lines or directions for vectors; and since a pair of vectors defines a plane (in any dimension) the bivectors are planes, whilst a trivector is a volume (and so on), and the scalar a point/number. Thus one sees that despite considering the geometry of three dimensions, there is actually a natural eight-dimensional space associated with it. Furthermore, there is also a four-dimensional subalgebra, the even subalgebra consisting of the scalar and the bivectors (the one that satisfies quaternionic relations).

This subalgebra is in fact the 4D space that allows us to make a connection between the geometries of 3D and 4D, namely to define 4D root systems from 3D root systems in Section 4.4.1. We have also used the full 8D algebra in other work [21, 45], e.g., for constructing the root system E_8 from the icosahedron H_3 or for defining representations. The presence of the pseudo-scalar allows us to dualise the even and odd subalgebras; for instance, we can dualise vectors into bivectors using $I = e_1 e_2 e_3$, as shown in the equation above $e_1 e_2 = I e_3$. The only other Clifford algebra we will consider here is that of 4D in Section 4.4.2, which is analogous with highest grade element (pseudoscalar) $e_1 e_2 e_3 e_4$.

But for now we go back to 3D to elucidate the relationship with the quaternions. As we have seen $e_1 e_2$, $e_2 e_3$ and $e_3 e_1$ are imaginary units, and in fact since $e_1 e_2 e_2 e_3 = e_1 e_3 = -e_3 e_1$ they are isomorphic to the quaternions. There are thus many different instances in Clifford algebras where geometric objects satisfy quaternionic relations, such as any three orthogonal planes, but there are also other examples, such as $\{1, e_t, e_x e_y e_z, e_t e_x e_y e_z\}$ in the Clifford algebra describing Minkowski spacetime (though this will concern us no further here). There are thus many geometric objects that may satisfy quaternionic relations, and treating them all the same (or assuming they commute with each other) can lead to great confusion and loss of geometric insight.

In particular, it is easy to see how pure quaternions, i.e., quaternions without a scalar part, are essentially bivectors (e.g., $e_2 e_3$) and thus easily dualised to vectors if one has the pseudoscalar I at one's disposal:

$$e_2 e_3 \leftrightarrow I e_2 e_3 = e_1 e_2 e_3 e_2 e_3 = -e_1.$$

(This is essentially the analogue of Hodge duality in the exterior algebra.) 3D vectors in general and 3D root systems can thus have representations in terms of pure quaternions. However, a cautionary tale that this is not always the case when the inversion/pseudoscalar is not available is in [46]. For example, the tetrahedral root system does not contain the inversion and thus has no representation in terms of pure quaternions. We will look at the binary tetrahedral group $2T$ as a group of elements of a Clifford algebra (multivectors) in detail below, in particular this lack of the inversion.

For essentially the same reasons as for the quaternions one can define a 4D Euclidean norm on the space of 3D spinors $R = a_0 + a_1 e_2 e_3 + a_2 e_3 e_1 + a_3 e_1 e_2$

through the inner product $(R_1, R_2) = \frac{1}{2}(R_1\tilde{R}_2 + R_2\tilde{R}_1)$ which gives $|R|^2 = R\tilde{R} = a_0^2 + a_1^2 + a_2^2 + a_3^2$. We will use this in Section 4.4.

The double-sided action of a quaternion on a pure quaternion amounts to a rotation; great significance is usually attached to this, but from Section 3.4.4 we recognise this as accidentally isomorphic to the general double-sided action / the simple reflection formula in Clifford algebras. Spin(4) is also accidentally isomorphic to two copies of SU(2) such that in 4D a pair of quaternions can describe a rotation. However, this is very backwards and does not generalise, compared with the general Clifford versor formula. Many of the hallmarks of why quaternions are seen as magical are thus actually completely general properties in Clifford algebras.

We saw that the Platonic symmetries – i.e., the rotational symmetries as discrete finite subgroups of SO(3), and hence, its double cover, SU(2) – follow an ADE classification. Now, let us see this from the Clifford perspective.

3.4.5 Double Cover the Clifford Way

Concatenating reflections in the Clifford algebra is equivalent to multiplying together root vectors, with double-sided application a la Section 3.4.4, giving rise to binary double covers of the polyhedral groups. Rotations correspond to an even number of reflections such that a rotation is doubly covered by a general spinor $R = a_0 + a_1 I e_1 + a_2 I e_2 + a_3 I e_3$, which acts on a vector as $\tilde{R}xR$. The rotation matrix this corresponds to is

$$\frac{1}{2}\begin{pmatrix} a_0^2 + a_1^2 - a_2^2 - a_3^2 & -2a_0a_3 + 2a_1a_2 & 2a_0a_2 + 2a_1a_3 \\ 2a_0a_3 + 2a_1a_2 & a_0^2 - a_1^2 + a_2^2 - a_3^2 & -2a_0a_1 + 2a_2a_3 \\ -2a_0a_2 + 2a_1a_3 & 2a_0a_1 + 2a_2a_3 & a_0^2 - a_1^2 - a_2^2 + a_3^2 \end{pmatrix}.$$

It is obvious that $-R$ gives rise to the same rotation matrix, since all the entries are quadratic in the spinor coefficients. So this gives the doubly covering homomorphism in the Clifford guise. Conversely, multiplying together vectors using the inner product to construct the Clifford algebra yields the double cover.

EXAMPLE 3.32 As a detailed example, we look at $2T$ as a group of Clifford multivectors. The following set of 24 spinors gives a Clifford realisation of the binary tetrahedral group, where group multiplication is given by algebra multiplication [47]. This particular set is based on the choice of the A_3 simple roots as $\alpha_1 = \frac{1}{\sqrt{2}}(e_2 - e_1)$, $\alpha_2 = \frac{1}{\sqrt{2}}(e_3 - e_2)$ and $\alpha_3 = \frac{1}{\sqrt{2}}(e_1 + e_2)$. The binary polyhedral group is given by even products of these root vectors. For instance, multiplying the two root vectors α_1 and α_3 gives

$$\alpha_1\alpha_3 = \frac{1}{\sqrt{2}}(e_2 - e_1)\frac{1}{\sqrt{2}}(e_1 + e_2) = \frac{1}{2}(e_2 e_1 - e_1 e_1 + e_2 e_2 - e_1 e_2) = -e_1 e_2,$$

a 2-fold rotation in the $e_1 e_2$-plane. Conversely,

$$\begin{aligned}
\alpha_1\alpha_2 &= \frac{1}{\sqrt{2}}(e_2 - e_1)\frac{1}{\sqrt{2}}(e_3 - e_2) \\
&= \frac{1}{2}(e_2 e_3 - e_1 e_3 - e_2 e_2 + e_1 e_2) \\
&= \frac{1}{2}(-1 + e_1 e_2 + e_2 e_3 + e_3 e_1).
\end{aligned}$$

This generates a 3-fold rotation, as one would expect from the A_3 diagram (like $\alpha_2\alpha_3$).

In the following table, the group is partitioned into conjugacy classes for easy reference (compare with Section 3.4.3). In addition, including odd products of reflections would further yield the pin group of order 48, which doubly covers the full Coxeter group A_3.

However, note that the inversion I is not among the group elements. Therefore, no spinors are the $I = e_1 e_2 e_3$-dual to a root vector. This inversion I sending $x \to -IxI = -x$ is not contained in the group, which means that the tetrahedron – unlike the other Platonic solids – is not inversion-invariant. Its inversion is an upside-down tetrahedron, and not the original one. This makes clear why the A_3 root system cannot have a representation in terms of pure quaternions: the pseudoscalar is not available to dualise vectors to bivectors.

Conjugacy class	Group elements
1	1
1_	-1
4	$\frac{1}{2}(1 - e_1 e_2 + e_2 e_3 - e_3 e_1),\ \frac{1}{2}(1 - e_1 e_2 - e_2 e_3 + e_3 e_1),$ $\frac{1}{2}(1 + e_1 e_2 - e_2 e_3 - e_3 e_1),\ \frac{1}{2}(1 + e_1 e_2 + e_2 e_3 + e_3 e_1)$
4_	$-\frac{1}{2}(1 - e_1 e_2 + e_2 e_3 - e_3 e_1),\ -\frac{1}{2}(1 - e_1 e_2 - e_2 e_3 + e_3 e_1),$ $-\frac{1}{2}(1 + e_1 e_2 - e_2 e_3 - e_3 e_1),\ -\frac{1}{2}(1 + e_1 e_2 + e_2 e_3 + e_3 e_1)$
4^{-1}	$\frac{1}{2}(1 + e_1 e_2 - e_2 e_3 + e_3 e_1),\ \frac{1}{2}(1 + e_1 e_2 + e_2 e_3 - e_3 e_1),$ $\frac{1}{2}(1 - e_1 e_2 + e_2 e_3 + e_3 e_1),\ \frac{1}{2}(1 - e_1 e_2 - e_2 e_3 - e_3 e_1)$
4$_-^{-1}$	$-\frac{1}{2}(1 + e_1 e_2 - e_2 e_3 + e_3 e_1),\ -\frac{1}{2}(1 + e_1 e_2 + e_2 e_3 - e_3 e_1),$ $-\frac{1}{2}(1 - e_1 e_2 + e_2 e_3 + e_3 e_1),\ -\frac{1}{2}(1 - e_1 e_2 - e_2 e_3 - e_3 e_1)$
6	$\pm e_1 e_2,\ \pm e_2 e_3,\ \pm e_3 e_1$

3.5 Lie Groups and Algebras

So far, we have encountered polyhedra, finite and continuous groups, root systems, finite graphs and binary polyhedral groups. As promised, we now introduce some rudiments of Lie theory. While the concept of root systems and Coxeter groups defined in Section 3.2.1 require no more than Euclidean geometry, it is well known that the motivation for their emergence comes from the study of **Lie groups**.

Now, such a topic as Lie groups is customarily reserved for a post-graduate course in mathematics, after some introduction to continuous transformations and differential geometry. We will here take an orthogonal approach and introduce Lie algebras straight away and leave the advanced geometry untouched, save for a few cursory remarks which we now make. Standard expositions of differentiable manifolds via charts and atlases can be found in the introductory pages of [48–50].

3.5.1 Lie Groups: A Lightning Introduction

Lie groups are simply continuous groups. Much like the finite, discrete rotation groups which we studied in detail for the Platonic solids, we are already familiar with a few Lie groups, but just perhaps not with a standardised notation. The most familiar is the set of rotational symmetries of $\mathbb{R}^3$, i.e., the set of orthogonal 3×3 real matrices. This easily generalises to $\mathbb{R}^n$:

$$\mathrm{O}(n) := \left\{ M \in \mathrm{Mat}_n(\mathbb{R}) \; : \; M^T \cdot M = \mathbb{I}_n \, . \right\} \tag{3.20}$$

Throughout, we will use $\mathrm{Mat}_n(\mathbb{F})$ to mean $n \times n$ matrices over $\mathbb{F}$, with $\mathbb{F}$ being $\mathbb{R}$ and $\mathbb{C}$ typically. We can also restrict to orientation-preserving rotations by imposing that the determinant is positive, these are the special orthogonal groups: $\mathrm{SO}(n) := \{M \in \mathrm{O}(n) \; : \; \det(M) = +1\}$.

Immediately, we have the so-called **Classical Lie groups**:

$$\text{Special Orthogonal } \mathrm{SO}(n) := \left\{ M \in \mathrm{Mat}_n(\mathbb{R}) \; : \; M^T \cdot M = \mathbb{I}_n, \det(M) = +1 \right\},$$

$$\text{Special Unitary } \mathrm{SU}(n) := \left\{ M \in \mathrm{Mat}_n(\mathbb{C}) \; : \; \overline{M}^T \cdot M = \mathbb{I}_n, \det(M) = +1 \right\},$$

$$\text{Symplectic } \mathrm{Sp}(2n) := \left\{ M \in \mathrm{Mat}_{2n}(\mathbb{R}) \; : \; \overline{M}^T \cdot \Omega \cdot M = \mathbb{I}_n \right\}, \tag{3.21}$$

with $\Omega := \left(\begin{smallmatrix} 0 & \mathbb{I}_n \\ -\mathbb{I}_n & 0 \end{smallmatrix} \right)$. We also mention, to set notation, the general linear group $\mathrm{GL}(n; \mathbb{R}) := \{M \in \mathrm{Mat}_n(\mathbb{R}) \; : \; \det(M) \neq 0\}$ of invertible matrices, as well as the special linear group $\mathrm{SL}(n; \mathbb{R}) := \{M \in GL(2; \mathbb{R}) \; : \; \det(M) = +1\}$. We can now say that finding the regular symmetries of $\mathbb{R}^3$ in Section 3.1 is the problem of classification of discrete, finite subgroups of SO(3). A key point is that in addition to these infinite families of classical groups (judiciously called types ABCD), there are also non-classical, or **Exceptional** Lie groups of type E, F, G. We are beginning to see our familiar pattern!

The fancier way of saying the above is that a Lie group is a compact differential manifold G which also possess a continuous group structure. For instance, our orthogonal group SO(3) is topologically (in fact diffeomorphically) the real projective space $\mathbb{RP}^3$ as a manifold, and so too, is SU(2), the three-sphere S^3 as a manifold. The reader is referred to [51–53] for this vast subject.

Because G has a group multiplication law, there is a distinguished point $\mathbb{I}_G$, serving as identity. The tangent space $\mathfrak{g} := T_{\mathbb{I}_G} G$ at this point is a vector space, further endowed with a bilinear product (an antisymmetric product known as the commutator), rendering $\mathfrak{g}$ an algebra. The group manifold G can be entirely reconstructed from $\mathfrak{g}$ (at least up to connected components). We recall that an algebra is just a vector space together with a bilinear product. More formally, we have

DEFINITION 3.33 A vector space $\mathfrak{g}$ is a Lie algebra if there is a (non-associative) bilinear form $\mathfrak{g} \times \mathfrak{g} \mapsto \mathfrak{g}$ called a Lie bracket $[\,,\,]$, with the properties

Anti-symmetry $[A, B] + [B, A] = 0$;
Jacobi identity $[A, [B, C]] + [B, [C, A]] + [C, [A, B]] = 0$

for elements $A, B, C \in \mathfrak{g}$.

The Lie bracket behaves like a commutator; indeed, $n \times n$ matrices, with operation $[A, B] = AB - BA$, form a Lie algebra. Thus, we say that a Lie algebra is abelian if $[A, B] = 0$ for all A, B.

Remark: Any vector space is defined over a ground field. In this book, we will exclusively consider the most common case of the Lie algebra as a vector space over $\mathbb{C}$.

For matrix groups, i.e., for subgroups of $GL(n; \mathbb{C})$, which is all that we consider in this book, to go from $\mathfrak{g}$ to G is simply matrix exponentiation.[23] Hence, we have a correspondence between G (the group is usually written in uppercase Italic, as in (3.21)) and the vector space $\mathfrak{g}$ (the algebra is usually written in lowercase Fraktur). One can easily check, using the exponentiation map (just checking to linear order in the Taylor series suffices), that for some of our above common Lie groups,

---

[23] As with any analytic function and any finite square matrix M, we can compute, via the formal Taylor series, $\exp(M) = \sum_{n=0}^{\infty} \frac{1}{n!} M^n$.

$$\text{General linear } \mathfrak{gl}(n;\mathbb{R}) = \mathrm{Mat}_n(\mathbb{R}),$$

$$\text{Special linear } \mathfrak{sl}(n;\mathbb{R}) = \{M \in \mathfrak{gl}(n;\mathbb{R}) \ : \ \mathrm{Tr}(M) = 0\},$$

$$\text{Special orthogonal } \mathfrak{so}(n) = \mathfrak{o}(n) = \left\{M \in \mathfrak{gl}(n;\mathbb{R}) \ : \ M + M^T = 0\right\},$$

$$\text{Unitary } \mathfrak{u}(n) = \left\{M \in \mathrm{Mat}_n(\mathbb{C}) \ : \ M + \overline{M}^T = 0\right\},$$

$$\text{Special unitary } \mathfrak{su}(n) = \{M \in \mathfrak{u}(n) \ : \ \mathrm{Tr}(M) = 0\},$$

$$\text{Symplectic } \mathfrak{sp}(2n) := \left\{M \in \mathrm{Mat}_{2n}(\mathbb{R}) \ : \ \Omega M = M^T \Omega\right\}, \tag{3.22}$$

with Ω as before. Note that on the algebra level, $\mathfrak{so}(n)$ is the same as $\mathfrak{o}(n)$ because the traceless condition is automatically satisfied by the anti-symmetry.

EXAMPLE 3.34 The simplest example is the Lie group $\mathrm{U}(1)$. Topologically, this is just S^1, a unit circle in the complex plane. We can parametrise the elements as $z = \exp(i\theta)$, with $\theta \in \mathbb{R}$ (note that as a manifold the coordinates are only defined locally, and the periodicity condition $\theta \in [0, 2\pi)$ is global), satisfying $\bar{z}z = 1$, in accord with the definition[24] in (3.21).

The Lie algebra $\mathfrak{u}(1)$ is likewise as in (3.22). Here, we have 1×1 matrices over $\mathbb{C}$, i.e., $\zeta \in \mathbb{C}$ such that $\zeta + \bar{\zeta} = 0$. In other words, $\zeta = i\theta$ is purely imaginary, as it should. Thus, the exponentiation map takes the Lie algebra $\mathfrak{u}(1) \ni \zeta$ to the Lie group $\mathrm{U}(1) \ni z = \exp(\zeta)$.

3.5.2 Simple Lie Algebras over $\mathbb{C}$

Adhering to the theme of the book, let us delve directly into the classification of Lie algebras (and thus, via the exponential map, that of Lie groups[25]).

As with any classification problem, we need to identify atomic elements, such as the primes for the integers, or the simple groups for finite groups. Here, the situation is no different and the fundamental building blocks are "simple Lie algebras". There are various *Structure Theorems*[26] that decompose $\mathfrak{g}$ into **simple Lie algebras**, which are non-abelian and contain no nonzero proper ideals, so that a semi-simple $\mathfrak{g}$ is a direct sum of these, and any finite $\mathfrak{g}$ is a

---

[24] The "determinant" here is just the complex modulus, which is equal to 1; this degenerate case we call $\mathrm{U}(1)$ instead of $\mathrm{SU}(1)$.

[25] Hence, the classification of of Lie groups reduces to that of Lie algebras, and a problem in differential geometry is reduced to one in linear algebra.

[26] We will not delve into the details of these. For example, Levi's decomposition theorem gives the building blocks of $\mathfrak{g}$: any finite dimensional Lie algebra is a semi-direct product of a normal solvable ideal and a **semisimple** algebra. Similarly, Langland's decomposition says that a parabolic subgroup P of G is the product of semisimple, abelian, and nilpotent subgroups.

semi-direct product of semi-simple Lie algebras and something called solvable Lie algebras. In short, we only need to classify the simple Lie algebras over $\mathbb{C}$.

The remarkable result of Cartan, Chevalley, Dynkin, Weyl, et al., is that the "classical" (infinite families of) groups (called type ABCD) and 5 exceptional cases (called type EFG) are all there is.

THEOREM 3.35 *The simple Lie algebras $\mathfrak{g}$ over $\mathbb{C}$ and their associated compact Lie groups G are*

Classical		
Name	$\mathfrak{g}$	G
$A_{n \geq 1}$	$\mathfrak{sl}_{n+1}$	$SU(n+1)$
$B_{n \geq 2}$	$\mathfrak{so}_{2n-1}$	$SO(2n-1)$
$C_{n \geq 3}$	$\mathfrak{sp}_{2n}$	$Sp(2n)$
$D_{n \geq 4}$	$\mathfrak{so}_{2n}$	$SO(2n)$

Exceptional
$E_{6,7,8}$
F_4
G_2

Sketch proof: The heart of the proof (cf. [54]) of this theorem lies in Dynkin's and Cartan's graphical representation of $\mathfrak{g}$ and the subsequent reduction of the classification to the Diophantine inequality (1.3). In brief, one starts with the theorem of Cartan–Weyl which renders the vector space $\mathfrak{g}$ into a convenient basis consisting of (1) a maximal abelian subalgebra $\mathfrak{h} \subset \mathfrak{g}$ within which all elements commute and the dimension of which is called the *rank* (the "Cartan subalgebra"); and (2) the complement vector space, ordered as the eigenspaces of the eigenvalue problem

$$[H, T] = \alpha_H T \,, \quad H \in \mathfrak{h}, \tag{3.23}$$

which give rise to root lattices. The eigenvalues α_H are called **roots** and one important part of the structure theorem is that these are precisely the abstract roots described in Section 3.2.1, satisfying the axioms of crystallographic root systems. Furthermore, the number of *simple roots* (once introducing a notion of positivity) is precisely the rank of $\mathfrak{g}$. Classifying root lattices (and in particular the simple roots) for various ranks thus classifies the semi-simple Lie algebras $\mathfrak{g}$ over $\mathbb{C}$.

Dynkin Diagrams: As with root systems where the diagrammatic approach to the simple roots by Coxeter and Dynkin was highly conducive to further work, so too, we can represent the table in Theorem 3.35 as Dynkin diagrams. The difference from Figure 3.3 is that in the Lie context, we customarily use multiple lines rather than labelled edges. This is because in the above Lie theory setting (unlike in the root system context) the eigenspace problem (3.23) leads to a restriction on the length of roots relative to one another. This then also fixes the angle between different roots automatically. This means that in Dynkin diagrams it is customary to have directed edges indicating the relative lengths, which because of the relation between lengths and angles gives the angle and thus the label above the link in the Coxeter–Dynkin diagram. For the crystallographic root systems and thus Lie algebras, we therefore have two slightly different diagrammatic representations: the Coxeter–Dynkin diagrams with labels, or the Dynkin diagrams with multiple edges (denoting the angles indirectly).

These are the building blocks of the Dynkin diagrams.

$$90° \ \bullet \quad \bullet \quad 120° \ \bullet\!\!-\!\!\!-\!\!\bullet \quad 150° \ \boxed{} \quad 135° \ \boxed{} \tag{3.24}$$

The classification of simple Lie algebras via Dynkin diagrams is shown in Figure 3.9. Note the similarity as well as the differences between Coxeter–Dynkin and Dynkin diagrams. When we do not allow double and triple lines, the resulting Dynkin diagrams are called **simply-laced**, corresponding to Lie algebras/Lie groups/root systems of type ADE as shown in the left of Table 3.2.

Affine Lie Algebras: As with root systems, we can have an extended, or affine version, by addition of an extra node in the Dynkin diagram. This has a specific meaning on the algebra level.

A natural generalisation of Lie algebras $\mathfrak{g}$ is the so-called **affine** Lie algebra. This is an *infinite dimensional* algebra defined as the tensor product with the algebra $\mathbb{C}[t, t^{-1}]$ of Laurent polynomials in a formal variable t, and then centrally extended by a one-dimensional center $c \in \mathbb{C}$:

$$\widehat{\mathfrak{g}} := \mathfrak{g} \otimes \mathbb{C}[t, t^{-1}] \oplus \mathbb{C}c , \tag{3.25}$$

together with an affine bracket

$$[a \otimes t^n + \alpha c, b \otimes t^m + \beta c] = [a, b] \otimes t^{n+m} + \langle a|b \rangle n \delta_{m+n,0} c ,$$
$$a, b \in \mathfrak{g}, \ m, n \in \mathbb{Z}, \ \alpha, \beta \in \mathbb{C}, \tag{3.26}$$

where $[a, b]$ is the usual bracket in $\mathfrak{g}$, $\langle a|b \rangle$ is the Killing form in $\mathfrak{g}$ and δ is the Kronecker symbol enforcing $m + n = 0$.

The classification of affine Lie algebras proceeds as with the ordinary case, with the extra caveat of the central extension. It turns out that all that is needed is to place one extra node to the ordinary Dynkin diagrams, corresponding to the affine (imaginary) root. Focusing again on the simply-laced cases, these are the affine diagrams on the right of Table 3.2, where the affine node is marked with a circle. These are the affine ADE algebras, usually denoted as $\widehat{ADE}$.

Lie algebras are a vast subject. Simple and semi-simple as well as affine Lie algebras are interesting classes within this; but the field is much wider with various generalisations of the above such as the Kac–Moody algebras, as well as others. In short, these define more general Lie algebras through generators and relations given by a generalised Cartan matrix, and many aspects of the above algebraic structure theory – such as root systems and representation theory – carry over to this Kac–Moody setting. The interested reader is directed to the literature [26, 28, 53] for further details and comprehensive treatments.

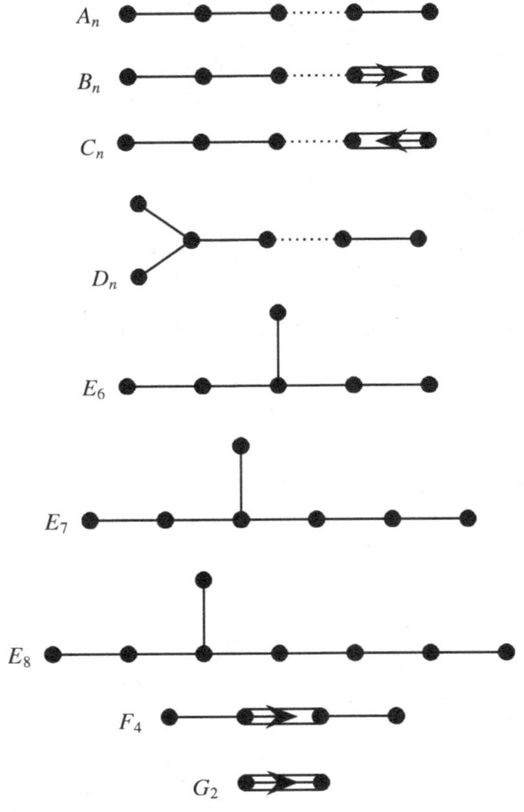

Figure 3.9 Classification of semi-simple Lie algebras via Dynkin diagrams. As is standard, the arrows in for the multiply connected edges (in the BCFG cases) are such that they point from longer to shorter roots. Since this is a book on ADE, we will not delve much into a discussion on these multi-edges.

3.5.3 Happy Accidents

Examining the Dynkin diagrams allows us to identify isomorphisms amongst some low-lying cases. These are happy accidents (law of small numbers) which can give very interesting properties (a part of so-called exceptional isomorphisms). It is worth tabulating them:

PROPOSITION 3.36 *At the level of Lie algebras, the following are isomorphic:*

$$A_1 \simeq B_1 \simeq C_1 \ i.e., \ \mathfrak{sl}_2 \simeq \mathfrak{su}_2 \simeq \mathfrak{so}_3 \simeq \mathfrak{sp}_1,$$
$$B_2 \simeq C_2 \ i.e., \ \mathfrak{so}_5 \simeq \mathfrak{sp}_4,$$
$$A_1 \times A_1 \simeq D_2 \ i.e., \ \mathfrak{su}_2 \oplus \mathfrak{su}_2 \simeq \mathfrak{so}_4,$$
$$A_3 \simeq D_3 \ i.e., \ \mathfrak{su}_4 \simeq \mathfrak{so}_6.$$

We have now encountered various incarnations of ADE patterns in different contexts: polyhedra, root systems, finite groups, graphs, Lie groups and algebras, etc. The next step will be finding connections between these different sets: the ADE correspondences. Some of these will seem straightforward, or even trivial, whilst others expose profound connections between vastly different areas of mathematics. The links that are still missing from this web of connections between ADE sets hint that there may be a lot more interesting mathematics to be developed and understood, in the guise of novel ADE correspondences. We hope that our readers are inspired to investigate these fascinating possibilities of discovery.

Exercises

3.1 Consider the following sets of three simple roots $\alpha_1, \alpha_2, \alpha_3$. Calculate the Cartan matrix and draw the Coxeter–Dynkin diagram in each case. According to the classification, which root systems do they represent?

(a) $\alpha_1 = (1, 0, 0)^T, \alpha_2 = (0, 1, 0)^T, \alpha_3 = (0, 0, 1)^T$

(b) $\alpha_1 = \frac{1}{\sqrt{2}}(-1, 1, 0)^T, \alpha_2 = \frac{1}{\sqrt{2}}(0, -1, 1)^T, \alpha_3 = \frac{1}{\sqrt{2}}(1, 1, 0)^T$

(c) $\alpha_1 = (0, 1, 0)^T, \alpha_2 = -\frac{1}{2}(\tau, 1, (\tau - 1))^T, \alpha_3 = (0, 0, 1)^T$

3.2 The E_8 root system is given by the following Coxeter–Dynkin diagram.

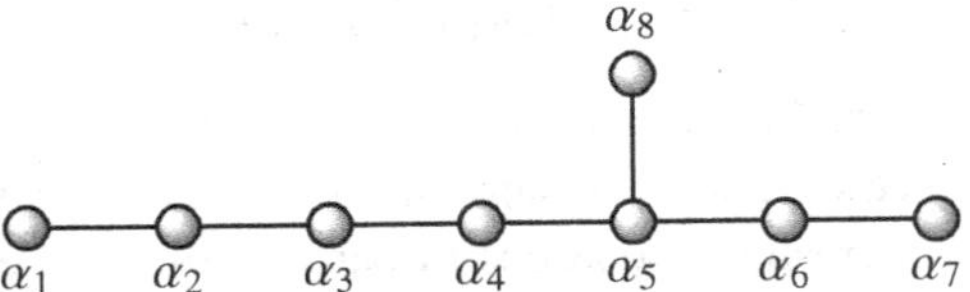

For the fundamental reflections s_i in the E_8 diagram above, and defining the pairwise combinations of generators

$$r_1 = s_1 s_7, r_2 = s_2 s_6, r_3 = s_3 s_5, r_4 = s_4 s_8$$

(as in Figure 3.7), show that these four new generators r_i themselves satisfy Coxeter group relations (you may assume that $(r_3 r_4)^5 = e$). Draw the corresponding Coxeter–Dynkin diagram.

3.3   The characteristic polynomial of an E_8 Coxeter element is given by

$$\lambda^8 + \lambda^7 - \lambda^5 - \lambda^4 - \lambda^3 + \lambda + 1.$$

Confirm that this can be factorised into two factors $\lambda^4 + \tau \lambda^3 + \tau \lambda^2 + \tau \lambda + 1$ and $\lambda^4 + \sigma \lambda^3 + \sigma \lambda^2 + \sigma \lambda + 1$. The Coxeter elements of which group might you expect to have characteristic polynomials of that form?

3.4   Consider the three root systems given by the simple roots

$$\alpha_1 = (1,0,0,0), \alpha_2 = (0,1,0,0), \alpha_3 = (0,0,1,0), \alpha_4 = \frac{1}{2}(-1,-1,-1,1);$$

$$\alpha_1 = \frac{1}{\sqrt{2}}(1,-1,0,0), \alpha_2 = \frac{1}{\sqrt{2}}(0,1,-1,0), \alpha_3 = (0,0,1,0), \alpha_4 = \frac{1}{2}(-1,-1,-1,1);$$

$$\alpha_1 = \frac{1}{2}(\tau,-1,0,\tau-1), \alpha_2 = (0,1,0,0), \alpha_3 = \frac{1}{2}(1-\tau,-1,-\tau,0), \alpha_4 = (0,0,1,0).$$

Which diagrams do these root systems have and what are the symmetries of these diagrams? What are the numbers of roots in these root systems?

3.5   A partial character table for the chiral icosahedral group $I = \mathfrak{A}_5$ is given here.

A_5	1	$12C_5$	$12C_5^2$	$20C_3$	$15C_2$
1					
3		τ	σ		
3'		σ	τ		
4					
5					

Partial character table for $\mathfrak{A}_5$.

Complete the character table for $\mathfrak{A}_5$.

3.6 Prove that a positive semi-definite symmetric $n \times n$ matrix of rank r is the matrix of inner products of a set of n vectors in $\mathbb{R}^r$. (See Theorem 5.4.)

3.7 Find the formula for the reflection of a vector $x = x_{\parallel} + x_{\perp}$ in the hypersurface defined by its unit normal vector n, where $x_{\parallel}$ is the component of x parallel to n, and $x_{\perp}$ is the orthogonal component (you can assume there is an inner product on the vector space), thus deriving formula 3.5.

3.8 Find the matrix that encodes such a reflection in the hyperplane with the unit normal vector $n = (n_1, n_2, n_3)^T$.

3.9 Consider the Clifford algebra with the new product ab given by $ab = a \cdot b + a \wedge b$. Using the symmetry resp. antisymmetry of the usual scalar product $a \cdot b$ resp. the wedge product $a \wedge b$, show that the symmetric resp. antisymmetric parts are related to these as $a \cdot b = \frac{1}{2}(ab + ba)$ resp. $a \wedge b = \frac{1}{2}(ab - ba)$.

3.10 Using the Clifford expression for the inner product, show that the reflection formula (3.5) simplifies to $x \to -nxn$. How are the reflections encoded in this way by n and $-n$ related? What is the inverse operation to such a reflection?

3.11 Show that Clifford products of unit vectors (where by "unit" we allow ± 1) form a group under the algebra product. Is it commutative? What are the inverses?

3.12 Consider the Clifford algebra of three-dimensional space, which is generated by the 3 orthogonal unit vectors e_1, e_2 and e_3. List the 3D root systems. What is the order of the Coxeter groups they generate? What is the order of their rotational subgroups? What is the order of the corresponding groups of products of (unit) root vectors in the Clifford algebra? What is the order of the binary double cover of the rotations? List the 4D root systems and identify which 3D root systems are related to which 4D root systems via these respective orders of the binary polyhedral groups.

3.13 Consider a set of bivectors $A, B, \ldots$ with the "commutator product" $A \times B = \frac{1}{2}(AB - BA)$. Show that this product is antisymmetric and satisfies the Jacobi identity (thus showing that this set of bivectors forms a Lie algebra). For two successive rotations in planes B_1 and B_2 via a rotor $R_i = e^{-B_i/2}$ show that the resulting rotor $R = e^{-B/2}$ describing the composite rotation has $B = B_1 \times B_2 +$ higher-order terms.

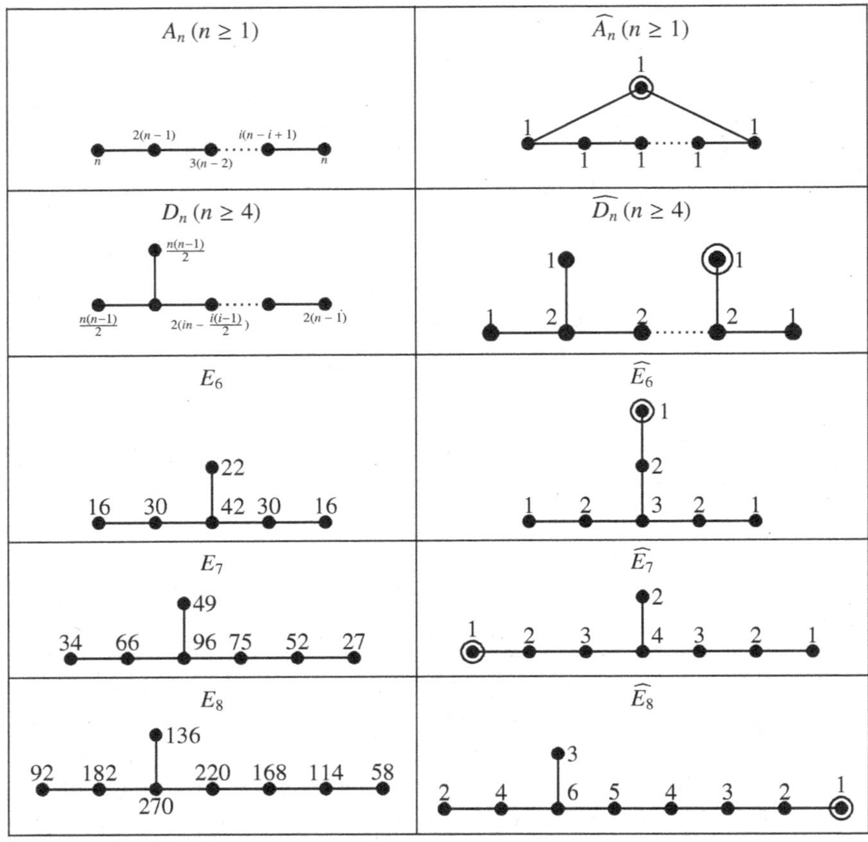

Table 3.2　*The eigenvectors of the ADE and extended (affine) ADE diagram adjacency matrix.*

The integer labels for the affine ADE diagrams are the Coxeter numbers (see also Sections 3.4.3, 4.3 and 4.5). For the ADE cases the eigenvectors are more complicated and have to do with the sum of the fundamental weights. Further details to the labels and coefficients of the highest root can be found in [34]. Simply put, the labels are such that: (1) for the affine ADE diagrams on the right, the label of each node is half of the sum of those adjacent to it; (2) for the ADE diagrams on the left, the label of each node is half the sum of the labels of those adjacent to it, subtracted by 2. In the $D_{n\geq 4}$ diagram, the two pitchfork nodes (the two extremal horizontal ones) have labels $n(n-1)/2$ and the others have $2(in - i(i-1)/2)$ with i between 1 and $n-2$. We will return in Section 5.8.1 with some remarks about labelling graphs.

4

ADE Correspondences

Having introduced various sets of algebraic objects that fall into ADE patterns in the previous section, we now move onto the connections between these different patterns, called ADE correspondences. It is of course these connections between different subjects (and the puzzling lack of connections between others) that make this subject so rich and interesting. We start with perhaps the most tantalising of connections: the one between the "Platonic" symmetries and the ADE diagrams. It is perhaps also the most mysterious, in its simple suggestiveness, but lacking a very rigorous construction. We cover some straightforward connections building up onto the perhaps best-known, and no less mysterious, connection, the McKay correspondence in Section 4.5, and also use this section to motivate new ADE correspondences. Woven throughout this chapter – in the spirit of the creative process of mathematics – is a recent example of a new ADE correspondence by one of the authors (PPD): we will argue step by step how to complete a correspondence of Trinities – via motivating the inclusion of suitable infinite families – to novel ADE correspondences. We hope that readers of this book will be stimulated and inspired to search for further novel ADE connections.

4.1 Trinity Revisited

We have already alluded to the "trinity" of exceptionals several times. Since this is a small set, it can often serve as a good starting point for noticing a full ADE-type of correspondence [55].

We saw that there is a mysterious suggestive link between the Trinities (A_3, B_3, H_3) and (E_6, E_7, E_8): the Platonic solids (A_3, B_3, H_3) have corresponding characteristic triples of orders of rotations $(233, 234, 235)$ generated by pairs of generators of the Coxeter groups, or via the angles between the simple

93

3D		Rot	*ADE*		Legs
A_3		2, 3, 3	E_6	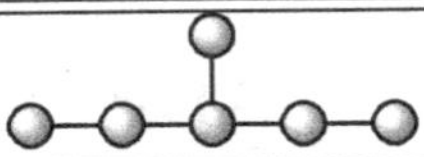	2, 3, 3
B_3	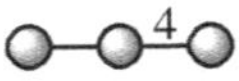	2, 3, 4	E_7	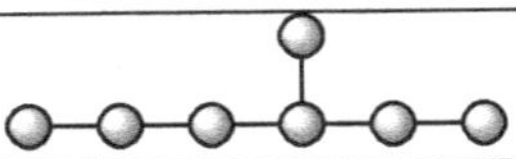	2, 3, 4
H_3	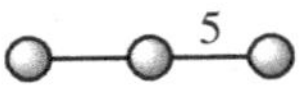	2, 3, 5	E_8	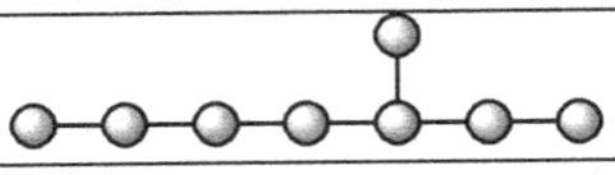	2, 3, 5

Table 4.1 *The orders of rotational symmetries from the Trinity* (A_3, B_3, H_3) *of Platonic root systems* $(233, 234, 235)$ *are equal to the lengths of the legs in the* (E_6, E_7, E_8) *diagrams.*

roots (as fractions of π). For instance, the icosahedral group has 2-, 3- and 5-fold rotations. The E-type diagrams also encode triples, albeit in a less obvious way: all three diagrams can be considered as consisting of three legs starting from a central node. The number of nodes in each leg also gives a number such that the three legs together give a triple. Thus, the E_8 diagram leads to the triple 235; likewise E_6 gives 233 whilst E_7 gives 234. Therefore both Trinities give rise to the same set of triples, as shown in Table 4.1. (Note also that the affine ADE diagrams correspond to the triples of the tessellations mentioned in Section 3.1.2).

This connection between Trinities seems puzzling but is well-known (see, e.g., [55]) and intuitive. It would be interesting to find a construction that makes this elusive link explicit and to see whether this observation could be extended to a full ADE correspondence. To which root systems would the A and D diagrams, i.e., the two countable families, correspond?

We will revisit this question soon in Section 4.6 after some detours, including first a connection between Platonic symmetries and 4D root systems (aka even discrete subgroups of SU(2)) in Section 4.4 and second a connection between discrete subgroups of SU(2) and the ADE Lie algebras (the McKay correspondence) in Section 4.5. This will motivate how to extend the above correspondence from just the Trinity to a full ADE correspondence by including the right two infinite families.

4.2 The Platonic Symmetries

The web of connections around the polytopes known as the Platonic solids and various related objects seems pretty obvious, almost trivial. For example, we have the links between the solids and their various symmetry groups: they have rotational symmetries, which are subgroups of SO(3) and give sets (mostly triples) of rotational symmetry orders (e.g., $(2, 3, 5)$ for the icosahedron), as well as the corresponding binary polyhedral groups in SU(2) (e.g., $2I$) and the full polyhedral groups/Coxeter groups in O(3) (e.g., H_3). The latter are in turn related to root systems and the coefficients of the highest root (e.g., $(1, 2, 2, 3, 3, 4, 4, 5, 6)$), as well as invariants such as exponents and degrees of invariant polynomials. The former are related to the representation theory of finite groups (e.g., $(1, 2, 2, 3, 3, 4, 4, 5, 6)$).

We can also think of the Platonic solids as regular tessellations of the sphere in terms of regular polygons and the link with the Diophantine inequality and triples of numbers. This then has the obvious analogue of the regular tessellations of the plane and the connection with the Diophantine equality (e.g., $(2, 3, 6)$).

We have explained these sets and many of the connections in the preceding sections as part and parcel of the development of the theory, rather than thinking of them as "mini-ADE correspondences". But we will now see that various members of this tightly connected cluster of ADE sets in turn have interesting correspondences linking them to other areas of mathematics, such as finite group theory and Lie theory.

4.3 ADE and Affine ADE

We have seen in Section 3.2.1 that crystallographic root systems are compatible with lattices. We have also seen from a graph-theoretic point of view that the graphs with eigenvalues equal to 2 and less than 2 are closely related and are the ADE and affine ADE diagrams. Here we will quickly review the procedure of affinisation, of constructing the lattice from a crystallographic root system – essentially by translating the root system. Note that the affine root is usually minus the highest root, and is given as a linear combination of the simple roots with coefficients that might look familiar by now – they will reappear in Section 4.5 on the McKay correspondence. For example, the affine root for E_8 is $-\alpha_0 = 2\alpha_1 + 3\alpha_2 + 4\alpha_3 + 5\alpha_4 + 6\alpha_5 + 4\alpha_6 + 2\alpha_7 + 3\alpha_8$. See Table 3.1 for the other cases.

Both these crystallographic root systems and the affine root systems are of course linked to the corresponding ADE and affine ADE Lie algebras.

4.3.1 Lattices from Root Systems

The reflections in a finite Coxeter group all leave the origin of V fixed and hence essentially act on the surface of a sphere. This gave rise to a compact reflection group in earlier sections, acting, e.g., on a polyhedron such as a Platonic solid. Here we will consider *affine reflections*, i.e., reflections with respect to more general hyperplanes not necessarily passing through the origin. Note that successive reflections in two parallel hyperplanes with separation d result in a translation of length $2d$ in the direction normal to the hyperplanes (see Figure 4.1). Such group generators will make the group non-compact and can build up a lattice structure, if certain compatibility conditions hold (i.e., the group is of crystallographic type and translating by just the right amount).

DEFINITION 4.1 (Affine hyperplane)   Given a vector α in an n-dimensional Euclidean space V and $K \in \mathbb{Z}$, we define an *affine hyperplane* as

$$H_{\alpha,K} = \{\lambda \in V \mid (\lambda \cdot \alpha) = K\}. \tag{4.1}$$

Note that therefore $H_{\alpha,0}$ corresponds to a plane through the origin and perpendicular to α, i.e., corresponds to the original H_α of a finite Coxeter group.

DEFINITION 4.2 (Affine reflection)   The reflection at $H_{\alpha,K}$ is called an *affine reflection* and is given by

$$s_{\alpha,K}: \quad v \mapsto v - \left(\frac{2(v \cdot \alpha)}{(\alpha \cdot \alpha)} - K\right)\alpha = v + K\alpha - \frac{2(v \cdot \alpha)}{(\alpha \cdot \alpha)}\alpha. \tag{4.2}$$

Note that this adds $K\alpha$ relative to the familiar reflection formula (3.5). With this definition of an affine reflection, one can extend a Coxeter group using this affine reflection as an additional generator. If one chooses just the right affine reflection, this results in a nice mathematical structure that generates lattices – otherwise it would just densely fill space. The extra generator one chooses [56] is one corresponding to the affine root α_0 with $\alpha_0 := d - \alpha_H$, where d is a vector in a space of dimension corresponding to the number of root vectors including the affine root such that

$$(d|\alpha_i) = 0 \ \forall i = 1, \ldots, n, \text{ and } \alpha_H \text{ the highest root.} \tag{4.3}$$

One often uses the vector $d = \vec{0}$ such that the affine root is just minus the highest root.

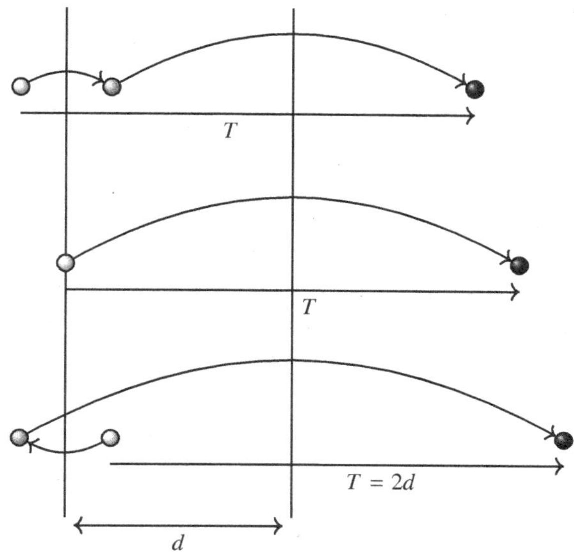

Figure 4.1 The result of the combination of two reflections in parallel hyperplanes is a translation by twice the separation of the hyperplanes: a starting point (white) is first reflected in the hyperplane on the left, resulting in the gray point; when this is reflected in the second hyperplane (right), it gives rise to a point (black) that is the translation of the original white point by twice the separation of the hyperplanes. This holds in each case, irrespective of where the starting point is in relation to the reflection hyperplanes. Thus including affine reflection generators essentially amounts to including a translation generator.

The extended group contains an additional generator, which effectively amounts to a translation obtained via one of the above combinations of reflections in parallel planes, i.e., the combination of an affine with a non-affine reflection with respect to the same root (see Figure 4.1). In particular, with

$$s_H^{\text{aff}} : v \mapsto (v + \alpha_H) - \frac{2(v \cdot \alpha_H)}{(\alpha_H \cdot \alpha_H)} \alpha_H, \tag{4.4}$$

we have

$$Tv = s_H^{\text{aff}}(s_H v)$$
$$= v + \alpha_H,$$

i.e., the two generators in combination act as a translation by the highest root α_H. This translation by the highest root builds up the lattice in each case, and for the (simply-laced) ADE cases this yields uniquely the simply-laced affine ADE cases.[1]

It is also worth noting that the somewhat miraculous collapse of the above formula is perhaps not so miraculous in Clifford algebra. The affine reflection formula there is

$$s_H^{\text{aff}} : v \mapsto -\alpha_H v \alpha_H + \alpha_H = -\alpha_H(v - \alpha_H)\alpha_H, \tag{4.5}$$

whilst the usual compact reflection is

$$s_H : v \mapsto -\alpha_H v \alpha_H \tag{4.6}$$

such that in combination they yield

$$s_H^{\text{aff}} s_H : v \mapsto -\alpha_H(-\alpha_H v \alpha_H)\alpha_H + \alpha_H = v + \alpha_H, \tag{4.7}$$

which is trivially a translation by the highest root. One could argue that Equation (4.2) amounts to translating the reflection plane, which already has a translation implicit anyway.

This constitutes our first relation between ADE sets: that of ADE diagrams and that of affine ADE diagrams. The latter can be achieved from the former via the affinisation procedure described above. The converse is rather simpler.

4.3.2 ADE from Affine ADE

Conversely, we have seen in Section 3.3 that the affine ADE diagrams have largest eigenvalue 2. The diagrams with largest eigenvalue < 2 have to be subgraphs of these, i.e., the ADE diagrams. This is because an induced subgraph has an adjacency matrix which is a principal submatrix of that of $\mathcal{G}$, so its greatest eigenvalue cannot exceed that of $\mathcal{G}$ (cf. Smith's Theorem 3.9). Thus one gets the ADE diagrams from the affine ADE diagrams.

Alternatively, going from affine ADE to ADE at the Coxeter group or root system level is also straightforward: one simply considers the subgroup or sub root system one gets by deleting the node corresponding to the (generator corresponding to) the affine root, i.e., one drops to considering just the compact part. For crystallographic groups, the corresponding lattice is just achieved by having infinitely many copies of the same repeating unit – one can thus simply go from the lattice (affine ADE) to the root system describing the symmetries of the repeating unit (ADE).

---

[1] One can seek a suitable if obviously limited generalisation of the notion of affinisation to include the non-crystallographic groups and suitably generalised, new algebraic structures. More details are in [25, 57–59] for the interested reader.

4.4 3D to 4D Root Systems

We have all the background that we need in order to prove that any 3D reflection group/root system yields a 4D reflection group/root system. The proof uses the Clifford algebra approach above (see Sections 3.2.1 and 3.4.4). The setting for a root system is a vector space with an inner product, so without loss of generality, one can consider this Clifford algebra over that vector space [16]. This induction theorem will give us something a bit more than a Trinity, and a bit less than an ADE correspondence – it has only one infinite family. We will motivate which infinite family could complete this to a full ADE correspondence.

4.4.1 Root System Induction via Clifford Algebra

Multiplying root vectors together in the Clifford algebra (e.g., of 3D space) generally yields pinors ('multivectors') in the full algebra (e.g., here 8D). Even products will in fact stay within the even subalgebra (e.g., here 4D). The corresponding rotational polyhedral group (a discrete subgroup of the special orthogonal group SO(3)) acts on a vector x via $\tilde{R}xR$ (a rotation). The spinors R themselves give a (spin) double cover of this polyhedral group, as both R and $-R$ encode the same rotation. These spinors themselves therefore form a group under multiplication $R_1 R_2$ in the algebra. This is of course the respective binary polyhedral group, which is a discrete subgroup of the spin group $\text{Spin}(3) = \text{SU}(2)$.

A general spinor in 3D thus has components in the even subalgebra

$$R = a_0 + a_1 e_2 e_3 + a_2 e_3 e_1 + a_3 e_1 e_2 \, , \tag{4.8}$$

which is a four-dimensional vector space. We saw above in Section 3.4.4 that we can also endow this vector space with a Euclidean inner product by defining $(R_1, R_2) = \frac{1}{2}(R_1 \tilde{R}_2 + R_2 \tilde{R}_1)$ for two spinors R_1 and R_2. This induces the Euclidean norm $|R|^2 = R\tilde{R} = a_0^2 + a_1^2 + a_2^2 + a_3^2$. It is thus very natural to think of spinors as living in Euclidean four-dimensional space (this is of course also linked to the quaternions as explained in Section 3.4). One can of course consider reflections in this 4D space, and the relevant scalar product is given by the formula here.

In fact, not only can we think of each 3D spinor as a 4D vector, but a spinor group actually yields a collection of such vectors in 4D. It is easy to show that this collection of vectors actually satisfies the axioms of a root system (Section 3.2.1). Therefore there is a direct correspondence between 3D root systems and 4D root systems via the intermediary spinor groups.

THEOREM 4.3 (Induction theorem)　*Any 3D root system gives rise to a spinor group G which induces a root system in 4D.*

Proof　Even products of 3D root vectors yield a collection of spinors in the even subalgebra (which has in fact a group structure). It was shown above that one can think of this space of spinors effectively as a 4D vector space with the inner product $(R_1, R_2) = \frac{1}{2}(R_1\tilde{R}_2 + R_2\tilde{R}_1)$. One can thus reinterpret a collection of spinors (the spinor group) as a collection of 4D vectors. We now check the two defining axioms of a root system (Definition 3.1) for this collection Φ of 4D vectors R:

(i)　By virtue of the standard Clifford algebra result, Φ contains the negative of a root R because both R and $-R$ encode the same rotation (since spinors provide a double cover). Thus if R is in Φ, then so is $-R$; other scalar multiples do not arise because of normalisation to unity.

(ii)　One can take reflections with respect to the inner product (R_1, R_2) defined above. Using the formula for the inner product in the reflection formula, these reflections are given by $R'_2 = R_2 - 2(R_1, R_2)/(R_1, R_1)R_1 = -R_1\tilde{R}_2R_1$ (normalisation). But G is closed under group multiplication, multiplication by -1 and reversal since $-R$ encodes the same group transformation as R, and $\tilde{R}$ is its inverse. Thus $-R_1\tilde{R}_2R_1 \in G$ for $R_1, R_2 \in G$ by closure of the group under group multiplication, reversal and multiplication by -1. Therefore Φ is invariant under all reflections in the 4D vectors R and is thus a root system.

$\square$

This proof does not make reference to any specific root system in 3D, and therefore allows one to construct a 4D root system for any 3D root system. This goes some way towards explaining why four dimensions are particularly rich for root systems because of additional, exceptional root systems such as H_4. This can be though of as due to the accidentalness of this construction. If we did not already know that they existed, this constructive proof could be used to construct them. One can calculate each case explicitly showing how each 4D root system arises from the 3D root system; but it is also obvious from the order of the 3D groups involved and the number of roots in the known 4D root systems what the correspondences have to be (see Exercises):

$$(A_3, B_3, H_3) \text{ give rise to } (D_4, F_4, H_4). \tag{4.9}$$

Above are the "irreducible" 3D root systems, which give Arnold's Trinity (A_3, B_3, H_3), though other reducible root systems of course also exist in 3D and can be used to construct corresponding 4D root systems. The case A_1^3 gives

A_1^4, and more generally the countably infinite family $A_1 \times I_2(n)$ gives rise to $I_2(n) \times I_2(n)$.

Examples of Induction

For illustrative purposes, we provide some more detailed examples here.

EXAMPLE 4.4 (Quaternion group: A_1^3 to A_1^4)   The simplest root system A_1 is just a root and its negative, so three copies of A_1 are just given by three orthogonal unit vectors as the simple roots such as $\alpha_1 = e_1, \alpha_2 = e_2, \alpha_3 = e_3$. Free multiplication (essentially amounting to multiplying together reflections in these simple roots) of these yields the eight elements in the 3D Clifford algebra from Section 3.4.4 and their negatives. Restricting to even products one gets the unit elements in the even subalgebra $\pm 1, \pm e_1 e_2, \pm e_2 e_3, \pm e_3 e_1$. These are essentially the quaternion group, and can be written as a collection of 4D vectors as $(\pm 1, 0, 0, 0)$ and permutations thereof. When thought of as a collection of 4D vectors, one sees that they are just the root system A_1^4.

EXAMPLE 4.5 (Binary tetrahedral group, A_3 and D_4)   Starting with the tetrahedral root system A_3 by multiplying the simple roots, e.g., given by

$$\alpha_1 = \frac{1}{\sqrt{2}}(e_2 - e_1), \alpha_2 = \frac{1}{\sqrt{2}}(e_3 - e_2), \alpha_3 = \frac{1}{\sqrt{2}}(e_1 + e_2),$$

one gets a group of 24 even products. This is the binary tetrahedral group consisting of 8 elements of the form $(\pm 1, 0, 0, 0)$ and 16 of the form $\frac{1}{2}(\pm 1, \pm 1, \pm 1, \pm 1)$. As a collection of 4D vectors they form the D_4 root system.

EXAMPLE 4.6 (Binary octahedral group, B_3 and F_4)   The octahedral root system B_3, e.g., with a choice of simple roots

$$\alpha_1 = e_3, \quad \alpha_2 = \frac{1}{\sqrt{2}}(e_2 - e_3) \text{ and } \alpha_3 = \frac{1}{\sqrt{2}}(e_1 - e_2),$$

yields the root system F_4 via even products of roots, which form the binary octahedral group of order 48. They include the 24 spinors we just had for D_4 together with the 24 "dual" ones of the form $\frac{1}{\sqrt{2}}(\pm 1, \pm 1, 0, 0)$.

EXAMPLE 4.7 (Binary icosahedral group, H_3 and H_4)   Finally, the icosahedral root system H_3 gives rise to H_4 in 4D. Multiplying even numbers of root vectors one gets 120 spinors that form the binary icosahedral group, doubly covering the 60 rotations of A_5. For the choice of simple roots

$$\alpha_1 = e_2, \alpha_2 = -\frac{1}{2}(\tau e_1 + e_2 + (\tau - 1)e_3), \alpha_3 = e_3,$$

there are 8, 16, and 96 respectively of the forms $(\pm 1, 0, 0, 0)$, $\frac{1}{2}(\pm 1, \pm 1, \pm 1, \pm 1)$ and $\frac{1}{2}(0, \pm 1, \pm(1 - \tau), \pm\tau)$, which form the H_4 root system.

One might think that this induction could be extended to arbitrary dimensions. And while the closure property holds trivially (because of the general spin double cover property in Clifford algebras), it is usually impossible to define a sensible inner product, and one loses associativity, etc., which makes it impossible to show the second root system axiom. This construction is thus accidental to 3D, although one can prove some partial results for the full algebra in 8D (though the result is pretty trivial), the octonions, as well as for 2D root systems. These also seem somewhat trivial but we shall discuss them briefly nonetheless for reasons that will become clear eventually [60]:

EXAMPLE 4.8 (2D root systems are self-dual)   As we have seen above, the Clifford algebra of two orthogonal unit vectors e_1 and e_2 is 4-dimensional,

$$\underbrace{\{1\}}_{\text{1 scalar}} \quad \underbrace{\{e_1, e_2\}}_{\text{2 vectors}} \quad \underbrace{\{e_1 e_2\}}_{\text{1 bivector}},$$

with the spaces of vectors and spinors both being two-dimensional. There is therefore a canonical mapping between vectors $v = a_1 e_1 + a_2 e_2$ and spinors $R = a_1 + a_2 e_1 e_2 =: a_1 + a_2 I = e_1 v$ via multiplication, e.g., with e_1, which is a bijection. Thus, for instance, for the choice of simple roots $\alpha_1 = e_1$ and $\alpha_2 = -\cos\frac{\pi}{n} e_1 + \sin\frac{\pi}{n} e_2$ one sees that taking the spinors of this root system is essentially multiplication by e_1 and therefore the whole root system $I_2(n)$ gets dualised to itself.

The space of spinors has a natural Euclidean structure given by $R\tilde{R} = a_1^2 + a_2^2$, i.e., a two-dimensional Euclidean vector space. This induces a rank-2 root system from any rank-2 root system in a similar way to the 3D-to-4D construction above, but is rather less interesting, as it does not yield any new root systems but just maps $I_2(n)$ to $I_2(n)$.

One would thus conclude that this case is pretty dull! We shall see why it was worth discussing it.

Thus the induction theorem relates the above sets $(A_1 \times I_2(n), A_3, B_3, H_3)$ and $(I_2(n) \times I_2(n), D_4, F_4, H_4)$. They are not quite ADE sets – they are missing one of the infinite families! In order to make the full connection, we first reconnect with Arnold and the Trinity part, and the McKay correspondence, which motivate the inclusion of the other infinite family – which turns out to be the self-dual $I_2(n)$ root systems we have just discussed.

4.4.2 Arnold's Connection: Degrees and Exponents

Arnold had actually found an indirect link between the trinities (A_3, B_3, H_3) and (D_4, F_4, H_4) [2, 61], about which he says, "*Few years ago I had discovered an operation transforming the last trinity [(A_3, B_3, H_3)] into another trinity of Coxeter groups (D_4, F_4, H_4). I shall describe this rather unexpected operation later.*"

We shall follow this discussion cursorily here, though it is very roundabout: but it will motivate the following simple construction. Arnold decomposed the groups (A_3, B_3, H_3) in something he calls the Springer cone decomposition of Weyl chambers. The number and types of Springer cones decompose the orders of the polyhedral groups as

$$24 = 2(1 + 3 + 3 + 5),$$
$$48 = 2(1 + 5 + 7 + 11),$$
$$120 = 2(1 + 11 + 19 + 29). \tag{4.10}$$

He then noticed that these coefficients happen to be one less than the quasi-homogeneous weights of (D_4, F_4, H_4), which are $(2, 4, 4, 6)$, $(2, 6, 8, 12)$ and $(2, 12, 20, 30)$, respectively.

There is one immediate simplification: these numbers are actually more directly the exponents m_i of (D_4, F_4, H_4), which as we have seen in Section 3.2.4 are well known to be one less than the degrees of polynomial invariants [20]. That is, they are given by the complex eigenvalues $\exp(2\pi i m_i/h)$ of the Coxeter element and

$$d_i = m_i + 1$$

for degrees d_i. This streamlines Arnold's observation somewhat, but is still not much more than a suggestive link; in constrast, the above Clifford construction made this connection very explicit [46, 60, 62, 63] (see previous section).

In fact, this construction also contained the other cases in addition to just the Trinity, given by the infinite family $A_1 \times I_2(n)$. Via the induction theorem these give $I_2(n) \times I_2(n)$ on the other side of the correspondence. One can thus wonder whether Arnold's observation in terms of the decomposition of the groups and the exponents might extend to include these additional cases from an infinite family. Indeed, it turns out that they do, which further establishes the connection between $(A_1 \times I_2(n), A_3, B_3, H_3)$ and $(I_2(n) \times I_2(n), D_4, F_4, H_4)$. We will therefore look at the computations necessary to include this infinite family in Arnold's correspondence in the next subsection, which the reader may wish to skip.

4.4.3 Aside: Arnold's Observation Holds for Our Additional Cases

This calculation for the 2D cases is somewhat trivial, but involves finding the exponents (i.e., complex eigenvalues) of the Coxeter element (shown in Table 3.2). These additional cases turn out to have exactly the right exponents to match the same decomposition in terms of Springer cones. Therefore, Arnold's observation extends from merely the Trinity to the wider 3D to 4D induction case, i.e., including $A_1 \times I_2(n)$.

The actual calculation is simple, but it is instructive to look at the construction from Section 3.2.4 from a different point of view, using Clifford algebras (Section 3.4.4). The usual procedure for finding exponents is to complexify the vector space (which was assumed to be Euclidean in Section 3.2.1) and look for complex eigenvalues of the Coxeter element. Complex eigenvalues of course just mean that the vector is not an eigenvector at all, but gets rotated in some plane and only returns back to itself after several applications. So it is more natural to think of the plane as an eigenplane rather than of complex eigenvectors. This is in fact unnecessary in the Clifford algebra setting and masks the underlying geometry, as we will now show [64, 65]. In Clifford algebras, the (double cover of the) Coxeter element decomposes into bivector exponentials, with the bivectors giving different rotation planes, and the angle giving exactly the correct exponents. We now see that this gives rise to the correct eigenvalue equation.

EXAMPLE 4.9 (Clifford algebra rotations in the plane: $I_2(n)$) We look at the simplest example in Clifford algebra for how rotations in planes are described by bivector exponentials: we start by looking at the simplest root system examples, $I_2(n)$. Let us choose the simple roots $\alpha_1 = e_1$ and $\alpha_2 = -\cos\frac{\pi}{n}e_1 + \sin\frac{\pi}{n}e_2$. Successive reflections in these two vectors actually amount to an n-fold rotation given by $\tilde{W}xW$ with W given as

$$W = \alpha_1\alpha_2 = -\cos\frac{\pi}{n} + \sin\frac{\pi}{n}e_1e_2 = -\exp\left(-\pi e_1 e_2/n\right). \qquad (4.11)$$

Taking the nth power of this simply yields $\pm\exp\left(\pm\pi e_1 e_2\right) = \pm(-1)$ such that double-sided application of this, as is usual for a spinor, gives $+1$, as expected (this is because the Coxeter element has order n in $I_2(n)$). Here the bivector in the exponential is e_1e_2 – of course, describing the plane (it couldn't be anything else anyway). The angle in the exponential is $\pm\pi/n$, which gives n-fold rotations clockwise and counterclockwise via W and $\tilde{W}$, i.e., exponents 1 and $n-1$, as expected.

Thus no complexification is necessary, as the bivector e_1e_2 squares to -1 but gives the rotation plane algebraically [64, 65]. As we saw above, in Clifford algebras there are many objects that square to -1 (but can have non-trivial commutation relations).

EXAMPLE 4.10 (Complex eigenvalue equation in the plane)   Let us look at the above rotation again: $\tilde{W}vW$ with $W = \alpha_1\alpha_2 = -\cos\frac{\pi}{n} + \sin\frac{\pi}{n}e_1e_2$. We are interested in eigenvectors, i.e., $\tilde{W}vW = \lambda v$. So let us just look at the left-hand expression for now, and try to use the commutation relations in the Clifford algebra. We recall that the pseudoscalar in the plane $e_1e_2 = I$ anticommutes with vectors in the plane $v = ae_1 + be_2$:

$$vI = (ae_1 + be_2)e_1e_2 = ae_1e_1e_2 + be_2e_1e_2 = ae_2 - be_1e_2e_2 = ae_2 - be_1$$

whilst

$$Iv = e_1e_2(ae_1 + be_2) = ae_1e_2e_1 + be_1e_2e_2 = -ae_1e_1e_2 + be_1 = -ae_2 + be_1.$$

Therefore $Iv = -vI$. So in the above expression xW, if we want to commute $W = -\cos\frac{\pi}{n} + \sin\frac{\pi}{n}e_1e_2$ to the left of v, then the scalar part will commute, whilst the bivector part will anticommute, i.e., change sign. Thus W gets turned into $-\cos\frac{\pi}{n} - \sin\frac{\pi}{n}e_1e_2$. Switching the sign on this bivector part is actually exactly the same as reversal $\tilde{W}$ so we have that commuting W past v gives $vW = \tilde{W}v$.

So overall we get that

$$\tilde{W}vW = \tilde{W}\tilde{W}v = \tilde{W}^2v.$$

Plugging in the form of $W = -\exp(-\pi e_1e_2/n)$ now immediately gives a complex eigenvalue-type equation

$$\tilde{W}vW = \tilde{W}\tilde{W}v = \tilde{W}^2v = \exp(-2\pi e_1e_2/n)v$$

with complex eigenvalue $\lambda = \exp(-2\pi e_1e_2/n)$. Thus, the complex eigenvalue equation arises automatically here by commuting W past the v. There is no need to complexify this space to achieve this, as the complex structure is given by the bivector, and the exponent/eigenvalue is just given algebraically. Note, however, that this only holds when the vector lies in the rotation plane, which here is trivially the case. We will get different behaviour when it doesn't, as discussed below.

The Coxeter plane geometry of more interesting root systems is fascinating, and in particular the light that a Clifford algebra treatment sheds upon it. The reader is referred to the literature for details [47, 63, 64], but we briefly describe the construction relevant to the invariants involved in Arnold's observation.

Consider the Coxeter pinor $W = \alpha_1\ldots\alpha_n$ which gives the action of the Coxeter element as $\tilde{W}vW$. Assume that we can completely factorise it in terms of commuting factors $W = W_k \dots W_1$ with $W_i = \exp(\pi B_i m/h) = \cos(\pi m/h) + \sin(\pi m/h)B_i$. That is they are bivector exponentials of bivectors B_i describing orthogonal planes. The following two propositions govern the actions of each factor on vectors that are either in a plane or orthogonal to it.

PROPOSITION 4.11 (Vector orthogonal to the rotation plane)   *A vector v that is orthogonal to a plane described by the bivector B_i is invariant under rotation in that plane W_i.*

Proof   If a vector v is orthogonal to B_i then it is also orthogonal to two vectors that generate B_i upon multiplication. Hence it anticommutes with both, and therefore it commutes with B_i. So W_i also commutes with v such that $v \to \tilde{W}_i v W_i = \tilde{W}_i W_i v = v$. Since W and $\tilde{W}$ are inverses and cancel, the vector is invariant under W_i. $\square$

In contrast, when the vector lies in the rotation plane we recover our result from Example 4.10:

PROPOSITION 4.12 (Vector in the rotation plane)   *If a vector v lies in the plane B_i then rotation leads to a complex eigenvalue equation*

$$v \to \exp\left(\pm 2\pi m_i B_i/h\right)v.$$

Proof   If v lies in the plane defined by B_i, it will be orthogonal to one of the vector factors but parallel to the other. Thus, v now anticommutes with B_i. Commuting W_i through to the left now introduces a minus sign in the bivector part (but not the scalar part, see e.g., Equation 4.11). This is again equivalent to reversal $\tilde{W}_i$ and one therefore gets the usual complex eigenvector equation

$$v \to \tilde{W}_i v W_i = \tilde{W}\tilde{W}v = \tilde{W}_i^{\,2} v = \exp\left(\pm 2\pi m_i B_i/h\right)v$$

without the need for complexification, because the complex structure simply arises from the bivector describing the rotation plane. $\square$

THEOREM 4.13 (Coxeter elements and complex eigenvalue equations)   *Consider a Coxeter pinor W factorising into orthogonal eigenspaces $W = W_1 \dots W_k$ and a vector v lying in one of these planes B_i. Then the action of the Coxeter element wv reduces to the complex eigenvalue equation with respect to this plane B_i.*

Proof   Since they are orthogonal all the W_is commute with each other. Since v is orthogonal to all of them apart from B_i they therefore also commute with v by Proposition 4.11. They therefore all commute through and cancel out.

$$v \to \tilde{W}vW = \tilde{W}_k \dots \tilde{W}_1 v W_1 \dots W_k = \tilde{W}_1 W_1 \tilde{W}_2 W_2 \dots \tilde{W}_i v W_i = \tilde{W}_i v W_i.$$

Only W_i does not commute through since v lies in B_i, i.e., the one that defines the eigenplane that v lies in. But by Proposition 4.12 commuting W_i through is the same as reversal and therefore leads to the complex eigenvalue equation with respect to W_i

$$v \to \tilde{W}_i v W_i = \tilde{W}_i^{\,2} v = \exp\left(\pm 2\pi m_i B_i / h\right) v.$$

$\square$

Note that now all the exponents arise purely algebraically from the factorisation of the product of simple roots in the Clifford algebra. W_i and $\tilde{W}_i$ determine righthanded and lefthanded rotations in the respective eigenplanes giving rise to the pairs of exponents m_i and $h - m_i$.

EXAMPLE 4.14 (Exponents of H_4) For example, $\alpha_1 \alpha_2 \alpha_3 \alpha_4$ for H_4 factorises as $\exp\left(-\frac{\pi}{30} B_C\right) \exp\left(-\frac{11\pi}{30} e_1 e_2 e_3 e_4 B_C\right)$ for the Coxeter plane bivector B_C. This is constructed as described in Section 3.2.4 via the Perron–Frobenius eigenvector giving rise to a white vector and a black vector which together define a plane. In Clifford algebra, this plane is defined by the bivector B_C simply as the outer/exterior product of the black and white vectors (thus defining the plane they span). This gives exponents $1, 11, 19, 29$ directly from the simple roots without the need for complexification.

The complex structure in each exponential is given by the bivector of the respective plane. There are therefore several different complex structures, which emerge purely algebraically from the factorisation of the Coxeter element. Complex eigenvalues thus arise geometrically without the need to complexify the whole real vector space. Clockwise and counterclockwise rotations in the same plane trivially yield exponents m and $h - m$ corresponding to W and $\tilde{W}$, which are of course conjugate. This description in terms of Clifford algebra therefore yields much deeper geometric insight, whilst avoiding ungeometric and unmotivated complexification. One sees that the eigenvalues and eigenvectors are not so much eigenvectors with complex eigenvalues, but rather eigenplanes of the Coxeter element. The complex nature of the eigenvalue arises because bivector exponentials describe rotations in planes with the plane bivector acting as an imaginary unit. Like for the 2D groups, the 3D and 4D geometry is actually completely governed by the above 2D geometry in the Coxeter plane, since the remaining normal vector (3D) or bivector (4D) are trivially fixed. The Coxeter element acts as a rotation by $\pm 2\pi/h$ (clockwise and counterclockwise) in the Coxeter plane B_C, and in the plane defined by IB_C as h-fold rotations giving the remaining exponents algebraically.

	h	m_i	W
$I_2(n) \times I_2(n)$	n	$1, 1, n-1, n-1$	$\exp\left(\frac{\pi}{n} e_1 e_2\right) \exp\left(\frac{\pi}{n} e_3 e_4\right)$
D_4	6	$1, 3, 3, 5$	$\exp\left(-\frac{\pi}{6} B_C\right) \exp\left(\frac{\pi}{2} I B_C\right)$
F_4	12	$1, 5, 7, 11$	$\exp\left(-\frac{\pi}{12} B_C\right) \exp\left(\frac{5\pi}{12} I B_C\right)$
H_4	30	$1, 11, 19, 29$	$\exp\left(-\frac{\pi}{30} B_C\right) \exp\left(-\frac{11\pi}{30} I B_C\right)$

Table 4.2 *Clifford factorisations of the 4D Coxeter versors giving rise to the correct exponents (for a particular choice of simple roots).*

The treatment for the other 4D root systems is analogous. D_4 and F_4 can be shown to factorise in just the right way so as to give the correct exponents. Table 4.2 shows the factorisation for those groups, giving the invariants that Arnold noticed in his Springer cone decomposition. D_4, F_4 and H_4 were of course induced from the root systems A_3, B_3 and H_3 above, via the even subalgebra. For simplicity when discussing the Coxeter plane geometry above, rather than continuing to think of a 4D subspace of the 3D Clifford algebra, we switched to a formulation just in terms of the usual four Euclidean dimensions. But the whole Coxeter plane construction can also be explicitly performed in the even subalgebra [47]. Further details can also be found in [63] where such a factorisation has been performed for E_8 (which as we saw above is linked to H_4) in [64] as well as for other 4D root systems such as A_4 and B_4, and novel Clifford invariants of the Coxeter element have been considered in [66].

These cover the Trinity-part D_4, F_4 and H_4. Returning to our infinite families and extending Arnold's observation to these:

EXAMPLE 4.15 (Exponents of $I_2(n) \times I_2(n)$)   The two copies of $I_2(n)$ in $I_2(n) \times I_2(n)$ are orthogonal such that one gets two sets of 1 and $h - 1 = n - 1$, i.e., $(1, 1, n-1, n-1)$ as shown above. For example, take $\alpha_1 = e_1$, $\alpha_2 = -\cos\frac{\pi}{n} e_1 + \sin\frac{\pi}{n} e_2$, $\alpha_3 = e_3$, $\alpha_4 = -\cos\frac{\pi}{n} e_3 + \sin\frac{\pi}{n} e_4$ as simple roots, then

$$W = (-\exp(-\pi e_1 e_2/n))(-\exp(-\pi e_3 e_4/n)) = W_{12} W_{34} \,,$$

which is just two copies of the 2D case.

If we follow Arnold's reasoning, the order of the binary polyhedral group (here the dicyclic group coming from $A_1 \times I_2(n)$) should be decomposable in terms of the exponents of the corresponding 4D root system ($I_2(n) \times I_2(n)$) as

$$4n = 2(1 + 1 + (n-1) + (n-1)) = 4n.$$

Indeed this matches the correct invariants. This also actually matches Arnold's original Springer cone decomposition: for $I_2(n)$, the two bounding walls of a given Weyl chamber decompose the plane into a cone containing this Weyl chamber together with its backward cone, which also contains one Weyl chamber, whilst the complementary cones contain the other $n - 1$ Weyl chambers each. In the projective plane one therefore gets the decomposition $2n = 2(1 + (n - 1))$. For $A_1 \times I_2(n)$, one simply gets a doubling of this, since the A_1 just creates two copies of the $I_2(n)$ decomposition. One therefore gets the decomposition $4n = 2(1 + (n - 1) + 1 + (n - 1))$. This seems somewhat needlessly complicated!

This decomposition therefore matches the correct exponents of the Coxeter element even for the countably infinite family in the root system correspondence. Arnold's original link therefore extends to the full correspondence $(A_1 \times I_2(n), A_3, B_3, H_3) \to (I_2(n) \times I_2(n), D_4, F_4, H_4)$ between root systems. In the spirit of this book, we are halfway there to an ADE correspondence: we have added one infinite family to a Trinity. We will advocate the inclusion of a further infinite family in Section 4.6, using other connections in the ADE web such as the McKay correspondence, in order to fully complete this into an ADE set. Our hope with this book is that the readership will likewise be "snappers-up of unconsidered trifles" [67]; and that they will start making conjectures and proving new connections in this mathematical web for themselves, discovering a wealth of exciting new mathematics in the process.

We therefore introduce the McKay correspondence now – the original and mysterious link that opened the field of ADE.

4.5 The McKay Correspondence

We saw from Sections 3.4.2 and 3.4.3 that the discrete finite subgroups G of SU(2) follow an ADE pattern; likewise, we saw from Section 3.2.1 that the simply-laced root systems (i.e., those with simple roots at only 120-degree-angles) also follow this classification pattern – and hence also the simply-laced Lie algebras. A priori, these come from two different worlds, individually well-known by the beginning of the twentieth century. Interestingly, it was not until late in that century that a precise connection between the two was found [3].

Because G is a subgroup of SU(2), there is a distinguished complex (spinorial) representation (see Section 2.3), which we can denote as $\mathbf{2}$. These correspond to the 2×2 matrix generators of the group as presented in Equations (5.25) and (5.27). From an abstract point of view, when G is non-abelian, i.e., the D and E families, $\mathbf{2}$ is an irreducible representation. When G is abelian, i.e., A-family, $\mathbf{2}$ is a reducible representation, decomposition into a direct sum of a pair of conjugate one-dimensional irreducible representations: $\mathbf{2} = \mathbf{1} \oplus \mathbf{\bar{1}}$.

One then considers the following tensor product decomposition over all irreducible representations $\mathbf{r}_i$ of G:

$$\mathbf{2} \otimes \mathbf{r_i} = \bigoplus_j a_{ij}\mathbf{r_j}, \tag{4.12}$$

where $a_{ij} \in \mathbb{Z}_{\geq 0}$ denote the multiplicity in the decomposition.

The astute observation of [3] is to construct a graph representing this tensor product structure by performing the following:

- Let each irreducible representation denote a node;
- Let the multiplicity a_{ij} of the decomposition be the adjacency matrix of the finite graph;
- The result is the **McKay quiver** (a graph, see Section 3.3).

Of course, this can be done for any finite group with a chosen defining representation. The key, as the reader would have expected, is that for discrete, finite subgroups of SU(2) – the ADE binary polyhedral groups – something miraculous happens.

$\mathbb{Z}/(2\mathbb{Z})$: A Warm-up Example

It is expedient to illustrate how to obtain the McKay quiver from (4.12) for a simple case. Take, for example, $C_2 = \mathbb{Z}/(2\mathbb{Z})$, consisting explicitly of the 2×2 matrices

$$C_2 = \left\{ \begin{pmatrix} 1 & 0 \\ 0 & 1 \end{pmatrix}, \begin{pmatrix} -1 & 0 \\ 0 & -1 \end{pmatrix} \right\}. \tag{4.13}$$

We can check that this is an embedding of C_2 into SU(2) by checking the determinant and the unitarity. One can think of these matrices as actions on $\mathbb{C}^2$ with complex coordinates (u, v). The defining (fundamental) representation of this $\mathbb{Z}/(2\mathbb{Z})$ is thus the second element in Equation (4.13), i.e., the negative, $-\mathbb{I}_{2\times 2}$, of the identity matrix.

Abstractly, the group itself has two elements, which we can denote as $\{1, -1\}$. The irreducible representations of C_2 are, since it is abelian, all 1-dimensional. There are two of them:

(i) the trivial representation $\mathbf{1}$ where all group elements are represented by 1 (every finite group has this representation); and

(ii) the representation $\mathbf{1}'$ where where the two elements in Equation (4.13) are represented by 1 and -1 respectively.

Subsequently, we have 2 nodes of the McKay quiver, which can be labelled $\mathbf{1}$ and $\mathbf{1}'$.

The fundamental $\mathbf{2}$ in this case[2] is a direct sum $\mathbf{1}' \oplus \mathbf{1}'$ because it is the matrix $-\mathbb{I}_{2\times 2}$. Again, we emphasise that of all the ADE polyhedral groups, only type

---

[2] A non-SU(2) action such as $(u, v) \to (u, -v)$, for example, would correspond to $\mathbf{1} \oplus \mathbf{1}'$.

A (the cyclics) has this split of the fundamental **2** because it is abelian; for the others **2** is a faithful, irreducible, 2-dimensional defining representation.

Thus, for our present case (4.12) reads

$$(\mathbf{1}' \oplus \mathbf{1}') \otimes \mathbf{1} = \mathbf{1}' \oplus \mathbf{1}',$$
$$(\mathbf{1}' \oplus \mathbf{1}') \otimes \mathbf{1}' = \mathbf{1} \oplus \mathbf{1} . \tag{4.14}$$

With the full decomposition information, we can now construct the McKay quiver. There are two nodes, labelled by **1** and **1**′. For the node **1**, there are two arrows to node **1**′; likewise, for the node **1**′, there are two arrows to node **1**. Letting the pairs of bi-directional arrows be represented by a single line, the quiver looks like the diagram of an Oxygen molecule.

This is the Dynkin diagram for affine $\widehat{A}_1$.

In general, for $A_n = \mathbb{Z}/((n+1)\mathbb{Z})$, one gets a necklace of $n+1$ beads, which is the Dynkin diagram for affine $\widehat{A}_n$.

Character Tables

In general, we would like a systematic way of extracting a_{ij} given any finite group. To do so, we recall a little character theory for finite groups (e.g., we recall Definition 2.40 onwards).

DEFINITION 4.16 The character χ of a group element $g \in G$ is the trace of the d-dimensional matrix representation of g.

Thus we have the following properties (the first two are immediate):

- χ is constant over conjugacy classes of G since $\mathrm{Tr}(hgh^{-1}) = \mathrm{Tr}(g)$; however, it does depend on the representation, including its dimension. Hence, we can denote the character as

  $\chi_\gamma^{(i)}$: the character of the ith irreducible representation $\mathbf{r_i}$ for the conjugacy class γ.

As mentioned in Section 2.3, there are as many conjugacy classes as there are irreducible representations, say r.

- We can therefore organise the characters into a square table called the **character table** of G with (i) indexing the rows/irreducible representations and γ indexing the columns/conjugacies (with one extra row keeping track of the sizes of each conjugacy class); we denote the sizes of the γth conjugacy class as r_γ.

- For characters of direct sums and tensor products, one has that $\chi(g \oplus h) = \chi(g) + \chi(h)$ and $\chi(g \otimes h) = \chi(g)\chi(h)$. The fancier way to say this is that χ is a ring homomorphism from representations to $\mathbb{C}$, taking $(\oplus, \otimes) \to (+, \times)$.

- The most important thing about the character table is that it is – weighted by conjugacy class sizes – orthogonal:

$$\sum_{\gamma=1}^{r} r_\gamma \chi_\gamma^{(i)*} \chi_\gamma^{(j)} = |G|\delta^{ij} ; \tag{4.15}$$

here, $*$ is complex conjugation.

Thus armed, one can extract the matrix a_{ij} by taking the character of Equation (4.12), giving us

$$\chi_\gamma^2 \chi_\gamma^{(i)} = \sum_{j=1}^{r} a_{ij} \chi_\gamma^{(j)} , \tag{4.16}$$

with χ_γ^2 being the character for the defining **2** representation.

The orthogonality (4.15) allows us to readily invert Equation (4.12) to yield

$$a_{ij} = \frac{1}{|G|} \sum_{\gamma=1}^{r} r_\gamma \chi_\gamma^2 \chi_\gamma^{(i)} \chi_\gamma^{(j)*} , \tag{4.17}$$

which gives the entries of the adjacency matrix of the McKay quiver.

ADE McKay Quivers

Therefore, we need the $r \times r$ character tables of the discrete finite subgroups of SU(2) (cf. also Section 3.4.3). These are classically well known and for reference, we present them below. As customary, we add a first row which is the size of the conjugacy classes and add a first column, the naming of the irreducible representations:

Cyclic $\widehat{A}_n = C_{n+1} = \mathbb{Z}/((n+1)\mathbb{Z})$

	1	1	1	$\cdots$	1
Γ_1	1	1	1	$\cdots$	1
Γ_2	1	ϵ	ϵ^2	$\cdots$	ϵ^n
Γ_3	1	ϵ^2	ϵ^4	$\cdots$	ϵ^{2n}
$\vdots$	$\vdots$	$\vdots$	$\vdots$	$\cdots$	$\vdots$
Γ_n	1	ϵ^n	ϵ^{2n}	$\cdots$	ϵ^{n^2}

$$\epsilon = \exp\left(\frac{2\pi i}{n+1}\right) \tag{4.18}$$

Binary dihedral $\widehat{D}_n$

	1	1	2	$\cdots$	2	n	n
Γ_1	1	1	1	$\cdots$	1	1	1
Γ_2	1	1	1	$\cdots$	1	-1	-1
Γ_3	1	$(-1)^n$	$(-1)^1$	$\cdots$	$(-1)^{n-1}$	i^n	$-i^n$
Γ_4	1	$(-1)^n$	$(-1)^1$	$\cdots$	$(-1)^{n-1}$	$-i^n$	i^n
Γ_5	2	$(-2)^1$	$2\cos\frac{\pi}{n}$	$\cdots$	$2\cos\frac{\pi(n-1)}{n}$	0	0
$\vdots$	$\vdots$	$\vdots$	$\cdots$	$\vdots$	$\vdots$	$\vdots$	$\vdots$
Γ_{n+1}	2	$(-2)^{n-1}$	$2\cos\frac{\pi(n-1)}{n}$	$\cdots$	$2\cos\frac{\pi(n-1)^2}{n}$	0	0

$$(4.19)$$

Binary tetrahedral $\widehat{E}_6 = 2T$

	1	6	1	4	4	4	4
Γ_1	1	1	1	1	1	1	1
Γ_2	1	1	1	w^2	w^{-2}	w^{-2}	w^2
Γ_3	1	1	1	w^{-2}	w^2	w^2	w^{-2}
Γ_4	2	0	-2	-1	-1	1	1
Γ_5	2	0	-2	$-w^{-2}$	$-w^2$	w^2	w^{-2}
Γ_6	2	0	-2	$-w^2$	$-w^{-2}$	w^{-2}	w^2
Γ_7	3	-1	3	0	0	0	0

$$w = \exp\left(\frac{2\pi i}{3}\right) \quad (4.20)$$

Binary octahedral $\widehat{E}_7 = 2O$

	1	6	12	1	8	8	6	6
Γ_1	1	1	1	1	1	1	1	1
Γ_2	1	1	-1	1	1	1	-1	-1
Γ_3	2	2	0	2	-1	-1	0	0
Γ_4	2	0	0	-2	-1	1	$\sqrt{2}$	$-\sqrt{2}$
Γ_5	2	0	0	-2	-1	1	$-\sqrt{2}$	$\sqrt{2}$
Γ_6	3	-1	1	3	0	0	-1	-1
Γ_7	3	-1	-1	3	0	0	1	1
Γ_8	4	0	0	-4	1	-1	0	0

$$(4.21)$$

Binary icosahedral $\widehat{E}_8 = 2I$

	1	20	30	20	1	12	12	12	12
Γ_1	1	1	1	1	1	1	1	1	1
Γ_2	2	-1	0	1	-2	τ	σ	$-\sigma$	$-\tau$
Γ_3	2	-1	0	1	-2	σ	τ	$-\tau$	$-\sigma$
Γ_4	3	0	-1	0	3	σ	τ	τ	σ
Γ_5	3	0	-1	0	3	τ	σ	σ	τ
Γ_6	4	1	0	1	4	-1	-1	-1	-1
Γ_7	4	1	0	-1	-4	1	1	-1	-1
Γ_8	5	-1	1	-1	5	0	0	0	0
Γ_9	6	0	0	0	-6	-1	-1	1	1

$$\tau = \frac{1+\sqrt{5}}{2}$$
$$\sigma = \frac{1-\sqrt{5}}{2}$$

$$(4.22)$$

Thus, we sum over the columns of the above character tables and obtain $r \times r$ matrices a_{ij} from Equation (4.17). This was done in [3] and the matrices a_{ij} turn out to be precisely the adjacency matrices of the affine ADE Dynkin diagrams! That is,

THEOREM 4.17 *(McKay correspondence for* SU(2)*) The McKay quiver for the ADE discrete finite subgroups of* SU(2) *are the Dynkin diagrams of the corresponding affine ADE Lie algebra, in particular,*

- *The trivial one-dimensional irrep for the identity corresponds to the affine node;*
- *The matrix $m_{ij} = 2\delta_{ij} - a_{ij}$ is the Cartan matrix;*
- *The dimensions d_i of the irreps are the Coxeter labels of the root system in Section 3.2.1, and also the eigenvalues of the Smith graphs in Section 3.3.*

This last comment on Coxeter labels and irreducible representations implies that the Coxeter number of the ADE Lie algebra is equal to the sum of the dimensions of the irreducible representations of the binary polyhedral group, and that the sum of the squares of the labels is of course equal to the order of the binary polyhedral group.

$\widehat{E}_6$: An Illustration

Let us illustrate with the binary tetrahedral group of order 24. Straight-away, we see that the dimensions of the irreps $(1, 1, 1, 2, 2, 2, 3)$ are the coefficients of expansion of the affine root for $\widehat{E}_6$ in Table 3.1, as well as the eigenvalues of the $\widehat{E}_6$ graph in Table 3.2. The sum of their squares, of course, is 24, the size of the binary group.

Next, applying (4.17) to the character table (4.20), we find, associating the 7 nodes with the 7 irreps (labelling accoordingly) and distinguishing the trivial **1** with a circle

$$a_{ij} = \begin{pmatrix} & 1 & 2 & 3 & 2 & 1 & 2 & 1 \\ \hline 1 & 0 & 1 & 0 & 0 & 0 & 0 & 0 \\ 2 & 1 & 0 & 1 & 0 & 0 & 0 & 0 \\ 3 & 0 & 1 & 0 & 1 & 0 & 1 & 0 \\ 2 & 0 & 0 & 1 & 0 & 1 & 0 & 0 \\ 1 & 0 & 0 & 0 & 1 & 0 & 0 & 0 \\ 2 & 0 & 0 & 1 & 0 & 0 & 0 & 1 \\ 1 & 0 & 0 & 0 & 0 & 0 & 1 & 0 \end{pmatrix}$$

$$(4.23)$$

The Coxeter number of E_6 is 12, which matches the sum of the dimensions of the irreps of the binary tetrahedral group, whilst the squares of the labels of course sum to the order of the binary tetrahedral group, 24.

The eponymous McKay correspondence therefore established a profound link between two ADE sets – the binary polyhedral groups and the simply-laced semi-simple Lie algebras – and was the first example of an ADE correspondence. Taking dimensions of Equation (4.12) yields

$$2d_i = \sum m_{ij} d_j$$

and thus, with hindsight, the graphs we would expect are PF2; however, at the time this observation was very unexpected, and indeed there are more extensive links from subgroups of SU(2) to ADE via the Coxeter number, etc.

4.6 A Trinity of Correspondences?

Summarising the above sections, we have the following three-way (partial) correspondences. We have a Trinity of Platonic symmetries mapping onto the E-type Trinity. We have three exceptional cases and an infinite family of 3D root systems inducing even discrete subgroups of SU(2) / 4D root systems. And we have the seminal McKay correspondence between discrete subgroups of SU(2) and ADE algebras, which is the archetypal ADE correspondence.

We essentially have something between a Trinity and the McKay correspondence for 3D $\rightarrow$ 4D, since so far there is only one countably infinite family in the root system correspondence. However, the Trinity (A_3, B_3, H_3) is connected to the binary tetrahedral, octahedral and icosahedral groups $(2T, 2O, 2I)$ and the 4D root systems that they induce (D_4, F_4, H_4). In turn, in the McKay correspondence these are linked to (E_6, E_7, E_8) via the tensor product structure of the binary groups. This may be giving us some ideas for firstly completing an ADE set, and secondly finding another ADE correspondence!

4.6.1　$(24, 60, 120)$ **and** $(12, 18, 30)$

We saw from Section 4.5 that part of the McKay correspondence is that the sizes $(24, 60, 120)$ of the exceptional binary groups are the sums of the squares of the labels of the affine ADE diagrams (likewise for the A- and D-types: viz., $\sum_i^n 1 = n$ and $4 \times 1 + \sum_i^{n-1} 2^2 = 4n$). But what about just the sums themselves? These are the Coxeter numbers of the (non-affine!) Lie Algebras and in the classical SU(2) / affine ADE McKay correspondence they have no immediate interpretation or direct connection.

As always, we could begin, as guided by the exceptionals:

$$12 = 1 + 1 + 1 + 2 + 2 + 3,$$
$$18 = 1 + 1 + 2 + 2 + 2 + 3 + 3 + 4,$$
$$30 = 1 + 2 + 2 + 3 + 3 + 4 + 4 + 5 + 6. \tag{4.24}$$

We now note that $(12, 18, 30)$ is also the number of roots in (A_3, B_3, H_3), which hints at a direct correspondence between 3D root systems and ADE Lie algebras, making the suggestive link in Section 4.1 more concrete. Could we perhaps use this as guidance to extend the correspondence to a full ADE correspondence?

The root system $I_2(n) \times I_2(n)$ is essentially the dicyclic (or binary dihedral) group, which in the McKay correspondence is connected with the D_n series of Lie algebras. The sum of the dimensions of the irreps is $2n + 2$, whilst the Coxeter number of D_{n+2} is $2(n+1)$. In addition, $I_2(n) \times I_2(n)$ in our root system induction construction is linked with $A_1 \times I_2(n)$, suggesting a more direct link between $A_1 \times I_2(n)$ and D_{n+2}. The number of roots in this $A_1 \times I_2(n)$ root system is also $2n + 2$, matching the expected Coxeter number of D_{n+2} and suggesting that we are on the right track in extending the correspondence!

If we want to extend the root system correspondence even further, we therefore need one other countable family corresponding to A_n. This is in fact given by the 2D root systems $I_2(n)$. We had not previously included these in the correspondence, since 2D root systems are self-dual [60] as we discussed above and thus did not appear to give new results: as we saw above in Sections 3.4.4 and 4.4.1, in the Clifford algebra of 2D the spaces of vectors and spinors are both of dimension two and can be canonically mapped into each other. The 2D root systems are thus self-dual, i.e., just map $I_2(n)$ to $I_2(n)$. However, as a complex/quaternionic group, $I_2(n)$ is precisely the cyclic group of order $2n$,

2D/3D	$|\Phi|$	4D	G	$\sum d_i$	ADE	h
$I_2(n)$	$2n$	$I_2(n)$	C_{2n}	$2n$	A_{2n-1}	$2n$
$A_1 \times I_2(n)$	$2n+2$	$I_2(n) \times I_2(n)$	Dic_n	$2n+2$	D_{n+2}	$2(n+1)$
A_3	12	D_4	$2T$	12	E_6	12
B_3	18	F_4	$2O$	18	E_7	18
H_3	30	H_4	$2I$	30	E_8	30

Table 4.3 *The correspondence of three ADE sets at the level of root systems. The first is the connection between root systems in 2D/3D and 2D/4D via the Clifford spinor induction. The 4D root systems in turn are binary polyhedral groups related to the ADE-type root systems of the (affine) ADE Lie algebras via the eponymous McKay correspondence. Combining both correspondences thus extends to a correspondence between 2D/3D root systems and ADE-type root systems. Now the Coxeter number h of the ADE Lie algebras, the sum of the dimensions of the irreducible representations of the binary polyhedral group G, $\sum d_i$, and the number of roots $|\Phi|$ in the 2D/3D root systems are all tantalisingly the same.*

C_{2n}, which via the McKay correspondence is linked to the missing A_n family. The number of roots in $I_2(n)$ is $2n$, which matches with the Coxeter number $2n$ of A_{2n-1}. This therefore seems like a plausible additional infinite family to include in the correspondence. Including this family in the correspondence also matches Arnold's observation above, linking the decomposed group (e.g., in terms of Springer cones) with the exponents of the corresponding group (here it's just self-dual). This tentative connection is summarised in Table 4.3.

There seems to be some tension regarding only covering half of the A_n cases. This is because root systems are always even, whilst in the McKay correspondence the odd-order cyclic groups also correspond to A-type Lie algebras with even n which are therefore not covered. But for now let us assume that the correct ADE set is indeed $(I_2(n), A_1 \times I_2(n), A_3, B_3, H_3)$, which induces the root systems $(I_2(n), I_2(n) \times I_2(n), D_4, F_4, H_4)$.

Now we would like to use this to go directly from Platonic symmetries to ADE diagrams without the detour via the spinor group intermediaries – can we complete the observation from the Trinity in Section 4.1 about the rotational orders of the polyhedral groups and the legs in the ADE diagrams to a fully-fledged ADE correspondence?

Table 4.4 shows the diagrams one would get if one started with this set of root systems $(I_2(n), A_1 \times I_2(n), A_3, B_3, H_3)$ and constructed diagrams in the same way as in Section 4.1 where we went from (A_3, B_3, H_3) to (E_6, E_7, E_8)

2D/3D		Rot	ADE		Legs
$I_2(n)$	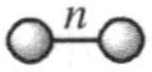	n	A_n	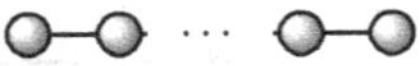	n
$A_1 \times I_2(n)$		$2, 2, n$	D_{n+2}	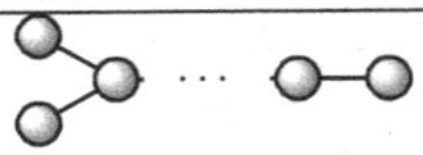	$2, 2, n$
A_3		$2, 3, 3$	E_6	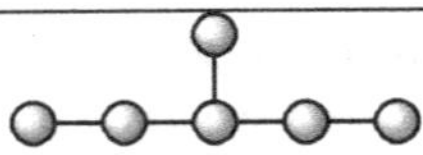	$2, 3, 3$
B_3	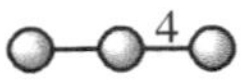	$2, 3, 4$	E_7	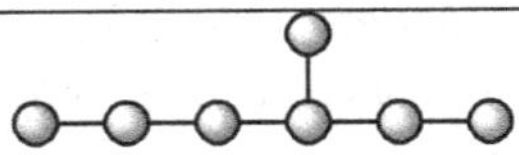	$2, 3, 4$
H_3	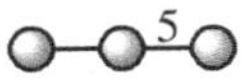	$2, 3, 5$	E_8	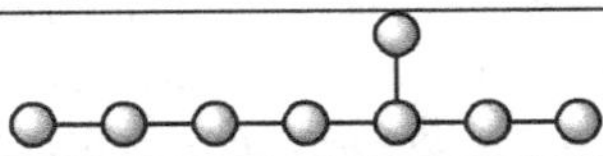	$2, 3, 5$

Table 4.4 *Extending the Trinity root system/E-type correspondence to a direct correspondence between 2D/3D and ADE root systems (i.e., omitting the intermediate step via 4D root systems/binary polyhedral groups).*
The 2D/3D root systems generate rotations of orders given by the angle between simple roots, i.e., given by the (implied) labels of the Coxeter–Dynkin diagrams. The rotation orders from the 2D/3D root systems are seen to be in a one-to-one correspondence with the lengths of the legs in the ADE Coxeter–Dynkin diagrams.

via the triples $(233, 234, 235)$ that are the orders of rotational symmetries in (A_3, B_3, H_3), which give precisely the lengths of legs of the E-type diagrams.

Extending this line of reasoning to the infinite families yields the following diagrams. The product of the two simple roots in $I_2(n)$ simply gives rise to an n-fold rotation. This corresponds to a single leg with n nodes in the diagram as a simple string and nothing else. This is exactly the A_n diagram. Actually here, in contrast with what was suggested above as regards the McKay correspondence, all A_n are achieved from the $I_2(n)$, and it is the A_n diagram, rather than the affine version $\tilde{A}_n$ that we get.

The root systems $A_1 \times I_2(n)$ encode a triple of rotational orders: the two simple roots of $I_2(n)$ still give a string of length n, but now the simple root of A_1 with either of the two simple roots of $I_2(n)$ will give rise to a 2-fold rotation, as they are orthogonal. The triple is therefore $22n$, which gives a diagram with one leg of length n which meets two legs of length 2, i.e., it is of D-type. It is in fact D_{n+2}. This was also to be expected from the connection with the McKay correspondence and exactly matches the number of roots, exponents, etc. This correspondence between rotation orders and lengths of legs in Dynkin diagrams therefore extends to the full set of root systems $(I_2(n), A_1 \times I_2(n), A_3, B_3, H_3)$ and the *ADE* Lie algebras.

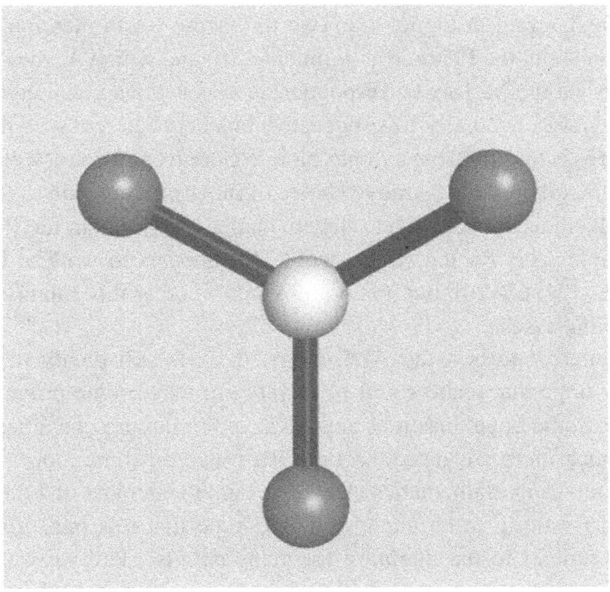

Figure 4.2 The D_4 graph (alternatingly coloured as in Section 3.2.4) symbolises the relationship between three different ADE classes of root systems and their three pairwise correspondences. This "trinity" of sets and connections might hint at a fundamental object at the centre yet to be identified.

A tangible construction connecting the two explicitly akin to the above construction in terms of Clifford spinors would be desirable, as this connection at the diagrammatic level seems a bit "hand-wavey". Our recent construction of the E_8 root system from the H_3 root system in a related Clifford construction [21, 22, 68] perhaps hints that this is again a correspondence between root systems, rather than operating at the level of the Lie algebras – after all, the Lie algebras and groups are defined by their respective root systems (up to integrability conditions which are met by crystallographic root systems). Table 4.4 summarises this correspondence.

This inclusion of the self-dual 2D root systems $I_2(n)$ therefore completes the three sets of *ADE* sets along with three *ADE* correspondences between them:

$$(I_2(n), A_1 \times I_2(n), A_3, B_3, H_3) \leftrightarrow (I_2(n), I_2(n) \times I_2(n), D_4, F_4, H_4)$$

$$\leftrightarrow (A_n, D_n, E_6, E_7, E_8).$$

There thus appears to be a web of connections between the three ADE sets of root systems that we first discussed in Section 3.2, with pairwise correspondences, and perhaps a central object connecting them all. This looks rather like the D_4 diagram in Figure 4.2. Two things are worth mentioning: there is still tension about the Platonic assignment $I_2(n)$ and which A_n they should correspond to via the McKay correspondence, versus the convincing assignment via the diagrams. Secondly, this suggestive link between root systems and ADE diagram legs is intuitive, but a rather more concrete connection would be desirable. Could perhaps some strange choice of inner product akin to the "reduced inner product" in terms of $\mathbb{Z}[\tau]$ integers map all the roots in the Platonic case to the simple roots for the ADE cases? This appears to work at least for H_3 going to E_8 [21, 68–70], but it is not obvious whether this can be extended to the remaining cases.

So despite progress using ADE-ology, there is still plenty to be puzzled about! We hope that readers will be inspired to make some progress in novel directions. Some open questions and advanced topics are contained in the following, much more advanced, section. We touch on some more ADE sets in advanced topics in mathematics as well as tantalising hints that there are deep connections waiting to be uncovered. We hope that this piques the readers' interest and defer to the literature for many details. Here we want to give a flavour of what the topics are about, to show a big picture of how they relate to each other and to ADE, and to arouse curiosity in our readers to investigate some of these open questions further.

Exercises

4.1 Find the orders n of the n-fold rotational symmetries generated by the root systems $(I_2(n), A_1 \times I_2(n), A_3, B_3, H_3)$. Draw Dynkin diagrams with legs of lengths of those rotational orders. Which diagrams do you get this way? What are their Coxeter numbers? What is the number of roots in the original root systems $(I_2(n), A_1 \times I_2(n), A_3, B_3, H_3)$?

4.2 By using the definitions, explicitly show that successive reflections in two parallel planes amount to a translation $s_H^{\text{aff}}(s_H v) = v + \alpha_H$. Think about why this might be rather more obvious.

4.3 Calculate the adjacency matrix for the tensor product of the binary tetrahedral group from Section 4.5, explicitly making the McKay connection with the $\tilde{E}_6$ diagram.

4.4 Consider the 3D root systems. What is the order of the Coxeter groups they generate, the order of their rotational subgroups and the order of the binary double cover of these rotations? Following the induction theorem 4.3 these generate 4D root systems. Confirm which 4D root systems get induced in this way. What is the sum of the irreducible representations of these binary polyhedral groups?

4.5 The binary polyhedral groups $2T, 2O, 2I$ are respective double covers of the polyhedral groups T, O, I. They have seven, eight and nine conjugacy classes, respectively, which essentially include those of the corresponding polyhedral group and also give rise to respective irreducible representations. That is, $2T, 2O, 2I$ have an additional three, three and four conjugacy classes, respectively, relative to T, O, I. What are the dimensions of the additional irreducible representations, given that they are all of even dimension? What is the sum of all the dimensions of irreducible representations in each case?

4.6 The E_8 root system is given by the following Coxeter–Dynkin diagram.

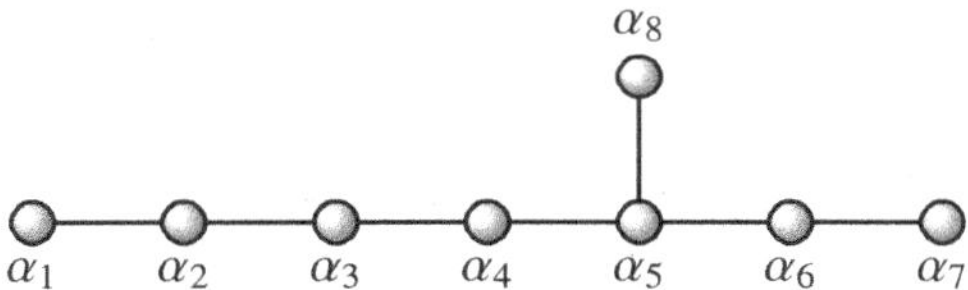

The E_7 and E_6 root systems are achieved by omitting α_1, and both α_1 and α_2, respectively. In terms of simple roots, the highest roots of E_6, E_7 and E_8 are given by

$$E_8 : \alpha_H = 2\alpha_1 + 3\alpha_2 + 4\alpha_3 + 5\alpha_4 + 6\alpha_5 + 4\alpha_6 + 2\alpha_7 + 3\alpha_8,$$

$$E_7 : \alpha_H = 1\alpha_2 + 2\alpha_3 + 3\alpha_4 + 4\alpha_5 + 3\alpha_6 + 2\alpha_7 + 2\alpha_8,$$

$$E_6 : \alpha_H = 1\alpha_3 + 2\alpha_4 + 3\alpha_5 + 2\alpha_6 + 1\alpha_7 + 2\alpha_8.$$

In each case one can use the highest root α_H as minus the affine root α_0 for an affine extension of the corresponding root system: $\alpha_0 = -\alpha_H$. Draw the extended Coxeter–Dynkin diagrams for these affine extensions of E_6, E_7 and E_8. The statement $\alpha_0 = -\alpha_H$ amounts to $\sum_{i=0...n} \alpha_i = 0$ for $n = 6, 7, 8$, respectively. Label each node in the extended diagrams by the corresponding coefficient in terms of the simple roots. What is the sum of the labels in each case? What is the sum of the squares of the labels in each case? What are the symmetry groups of the diagrams?

5

Advanced Miscellany

The preceding chapters have introduced the reader to a variety of cases where ADE meta-patterns arise, as well as provided the underlying reason for some of the correspondences between them. However, they touch but the surface of ADE-ology! This persistent trio – two infinite families and a triplet of exceptionals – emerges and still continues to emerge across the mathematical landscape.

In this last chapter, we will give the reader a glimpse of some of the more advanced topics, from finite group theory to geometry to physics, to representation theory, to number theory, etc., wherein we will encounter our familiar ADE friends. For the beginning students, this familiar theme will inevitably appear in unfamiliar territory, so we will unavoidably be more cursory in our presentation compared to the pedagogical efforts of the previous chapters. Nevertheless, we hope that a perusal might inspire them to further investigate this vast subject. The different topics in this section are largely independent from one another and can thus be perused at leisure. In particular, the section on Moonshine, as intriguing as it is, can be freely skipped, and the subsections are mostly only loosely connected.

5.1 Monstrous Connections

We now come to an advanced correspondence that is perhaps the least understood and most curious. Here, we do not seem to have a full ADE story yet and even for the trinity of exceptionals the connections are mysterious [55]. Nevertheless, the consequences are far-reaching enough that we open the present chapter with it.

First, we recall some results from the classification of finite simple groups. One can think of this as the discrete analogue of the classification of simple

Lie groups. Again, roughly, as integers are composed (via unique factorisation) from primes, Lie algebras (via structure theorems), from simple algebras, so are finite groups (via Jordan–Hölder composition) made from simple groups as building blocks. More precisely, a finite **simple group** is one which contains no non-trivial normal subgroup. Importantly, the composition series (chain of normal subgroups) of any finite group is unique up to permutations and isomorphism.

As is perhaps always the case, problems over the discrete are much harder: the classification of simple Lie algebras/groups followed quickly from that of the root diagrams, resulting from an Egyptian fraction type of Diophantine inequality. The classification of finite simple groups is much more difficult: it started with Galois and lasted until the recent volume of Gorenstein, spanning two centuries and many tens of thousands of pages of proof [71, 72]. We have

THEOREM 5.1 *A finite simple group is one of the following:*

- *a cyclic group C_p of prime order;*
- *an alternating group $\mathfrak{A}_{n \geq 5}$;*
- *the Lie groups defined over finite fields; and*
- *26 exceptional cases called the Sporadics, including various Mathieu, Conway and Fischer groups, the Monster and Baby Monster groups, as well as others.*

Remark: The first family is easy to understand: cyclic groups of prime order, by Lagrange's theorem, have no non-trivial subgroups at all.

The simplicity of the second family is what allowed Galois theory to prove the absence of finite radical solutions to polynomials of degree more than 4.

The third family is really a family of families: consider all our simple Lie groups from Section 3.5. Instead of having them defined over the complex numbers, consider them over finite number fields. For example, $\mathrm{SL}(2; \mathbb{C})$ is a Lie group which is the space of 2×2 complex matrices with unit determinant. If, instead, we consider $\mathrm{SL}(2; \mathbb{F}_2)$, with entries over the finite field of 2 elements, say ± 1, then the result is a finite group of size 6; in fact $\mathrm{SL}(2; \mathbb{F}_2) = \mathfrak{S}_3$. Similarly, we have that $\mathrm{SL}(2; \mathbb{F}_3) = 2T$, the binary tetrahedral group.

The simple groups of Lie type are divided up in various ways: for example, into classical groups (linear, symplectic, orthogonal, unitary) and exceptional groups (associated with the exceptional Lie algebras); or into Chevalley groups and twisted groups. See Carter [73] for further discussion of these groups.

Finally, the sporadics are completely mysterious. They are typically of enormous size, as we will soon see. There is much mathematics yet to be discovered, as these sporadics seem to be lying at the intersection between algebra, number theory, combinatorics and ADE-ology, connected to other exceptional mathematical objects.

Remark: Whilst the classification of finite simple groups is not in any obvious ADE pattern, it was an old speculation of McKay whether the largest three of the Sporadics (q.v. recent introduction in [55]), viz.,

$$\text{(Monster, Baby Monster, Fischer's Group)} \tag{5.1}$$

might furnish a trinity that relates to (E_6, E_7, E_8). We will expound upon this connection shortly.

5.1.1 Moonshine

To give an idea of how extraordinary the sporadic groups are, we begin with a brief recapitulation of Monstrous Moonshine, a correspondence between two seemingly utterly unrelated subjects:

(i) Finite group theory: Of the sporadic family of 26 finite simple groups, the largest is the so-called Monster group $\mathbb{M}$, of order

$$|\mathbb{M}| = 2^{46} \cdot 3^{20} \cdot 5^9 \cdot 7^6 \cdot 11^2 \cdot 13^3 \cdot 17 \cdot 19 \cdot 23 \cdot 29 \cdot 31 \cdot 41 \cdot 47 \cdot 59 \cdot 71 \sim 10^{54} \,. \tag{5.2}$$

It is remarkable that there should exist an outlying finite group of such astronomical size with no normal subgroups. Though large in size, there are only 194 conjugacy classes and irreducible representations. The first non-trivial irrep is of dimension 196883, being in the list

$$\{1, 196883, 21296876, 842609326, 18538750076, \ldots\} \,. \tag{5.3}$$

(ii) Number theory and modular forms: One of the central objects in number theory is an analytic object dating back at least to Klein, called the *j-invariant*. This is the "only" meromorphic function defined on the upper-half plane invariant under the full modular group $\Gamma := SL(2; \mathbb{Z})$; by "only" we mean that all invariant functions are rational functions in $j(z)$. Therefore the modular action of Γ on the upper half complex plane $\mathcal{H}$ fixes the field $\mathbb{C}(j)$ of rational functions of j. Other than a simple pole at $i\infty$, $j(z)$ is the only holomorphic function invariant under Γ once we fix the normalisation

$$j\left(\exp\left(\frac{2\pi i}{3}\right)\right) = 0 \,, \quad j(i) = 1728 \,, \qquad j(\gamma z) = j(z) \,, \gamma \in \Gamma \,. \tag{5.4}$$

Writing the **nome** $q := \exp(\pi i z)$, we obtain the famous Fourier expansion of $j(q)$ as

$$j(q) = \frac{1}{q} + 744 + 196884q + 21493760q^2 + 864299970q^3 + \cdots . \quad (5.5)$$

Now, the pole at $z = i\infty$ (i.e., for $q = 0$) is explicit and all the coefficients are positive integers. In the ensuing we will often make use of the *normalised* form where the constant 744 has been set to 0; this is habitually denoted as j_{M}; the subscript will become clear soon. Furthermore, we could divide by 1728 to ensure ramification only at 0, 1 and ∞. To clarify our convention, we adhere to the following

$$j_{arithmetic}(q) = j(q) = \frac{1}{q} + 744 + 196884q + 21493760q^2 + \cdots ,$$

$$j_{analytic}(q) = \frac{1}{1728} j(q),$$

$$j_{\mathrm{M}}(q) = j(q) - 744 . \quad (5.6)$$

The j-invariant is a special case of a **Hauptmodul**, or *principal modulus*. For a genus zero subgroup $\Theta \subset \Gamma$ where the quotient $\Theta\backslash\mathcal{H}$ is a Riemann sphere, its modular functions, i.e., the field of functions invariant under Θ, is generated by a *single function*, much like the aforementioned case of the full modular group $\Gamma = PSL(2, \mathbb{Z})$ where the j-invariant generates $\mathbb{C}(j)$. For higher genera, two or more functions are needed to generate the invariants, and there is not as nice a notion of a unique canonical choice.

Another McKay Correspondence

The initial observation by the last author, which started the subject of Moonshine, was that

$$196884 = 196883 + 1 , \quad (5.7)$$

where the left-hand side is the first term of the Fourier expansion of j_{M} in (5.6). This is so outlandish – that modular forms should relate to finite groups – it was deemed "moonshine" by Conway. Yet, a quick check of the next few coefficients of j_{M}, all of which can be written in simple sums of the form (5.7), led Conway and Norton to make a set of precise conjectures in [74], and Borcherds' Fields-Medal-winning proof [75].[1]

---

[1] The proof of the Monstrous Moonshine conjectures in terms of physics comes from compactifying chiral bosonic string theory on the 24-dimensional quotient torus of $\mathbb{R}^{24}$ by the Leech lattice Λ_{Leech} (which we will see more of later in this section) and orbifolded by a two-element reflection group, i.e., $\mathbb{R}^{24}/\Lambda_{Leech}/(\mathbb{Z}/2\mathbb{Z})$. The vertex operator algebra describing the resulting 2D CFT (the Monster vertex operator algebra) has the Monster as its automorphism group, as does its degree-2 piece, the Griess algebra, which is a commutative but non-associative algebra on a real vector space of dimension 196884. Borcherds applied the

We do not have space here to discuss the myriad of interesting connections in Moonshine but will only focus on its relations to ADE. For now, we only mention in passing that the idea behind the extraordinary fact in (5.7) is that there is an infinite-dimensional representation of $\mathbb{M}$

$$V = V_0 \oplus V_1 \oplus V_2 \oplus \cdots \tag{5.8}$$

with $V_0 = \rho_1$, $V_1 = \{0\}$, $V_2 = \rho_1 \oplus \rho_{196883}$, $V_3 = \rho_1 \oplus \rho_{196883} \oplus \rho_{21296876}, \ldots$, the corresponding generating function (graded dimension) of which is

$$\sum_{n=0}^{\infty} q^n \dim(V_n) = 1 + 196884q^2 + 21493760q^3 = q(j(q) - 744) = q j_{\mathbb{M}}(q). \tag{5.9}$$

In general, one has the generating function for the characters, known as a McKay–Thompson series, for each of the conjugacy classes of $\mathbb{M}$,

$$\begin{aligned}
T_g(q) &= q^{-1} \sum_{n=1}^{\infty} \mathrm{ch}_{V_n}(g) q^n \\
&= q^{-1} + 0 + h_1(g)q + h_2(g)q^2 + \cdots ,
\end{aligned} \tag{5.10}$$

and they all turn out to be modular functions (Hauptmoduln) of some group between $PSL(2; \mathbb{Z})$ and $PSL(2; \mathbb{R})$.

5.1.2 Monster and E_8

The initial observation that the j-function encodes the irreducible representations of $\mathbb{M}$, was part of another observation:

$$j(q)^{\frac{1}{3}} = q^{-\frac{1}{3}} \left(1 + 248q + 4124q^2 + 34752q^3 + \cdots \right) \tag{5.11}$$

encodes the irreducible representations of the Lie algebra E_8 (starting with $\{1, 248, 3875, 27000, 30380, 147250, \ldots\}$) in a similar fashion (note that $248 = 744/3$):

$$248 = 248, \quad 4124 = 3875 + 248 + 1, \quad 34752 = 30380 + 3875 + 2 \cdot 248 + 1 \ldots \tag{5.12}$$

This was in fact the first matter to be settled [76]: the unique level-1 highest-weight representation of the affine Kac–Moody algebra $E_8^{(1)}$ has graded dimension encoded by $j(q)^{\frac{1}{3}}$. One should also be mindful of the fact that the theta-series for the E_8 root lattice $\Lambda(E_8)$ is

$$\theta_{\Lambda(E_8)}(q) = \sum_{x \in \Lambda(E_8)} q^{|x|^2/2} = 1 + 240 \sum_{n=1}^{\infty} \sigma_3(n) q^n = E_4(q),$$

Goddard–Thorn theorem to construct a Monster Lie algebra, which is a generalised Kac–Moody algebra acted on by the Monster.

the Eisenstein series (where $\sigma_3(n)$ is the sum of the cubes of the divisors of n), so that we have

$$j(q) = \frac{\theta_{\Lambda(E_8)}(q)^3}{\Delta(q)} , \qquad \Delta(q) = \eta(q)^{24} , \tag{5.13}$$

where $\eta(q)$ is the Dedekind eta-function and $\Delta(q)$ is the Ramanujan delta-function. Other roots of $j(q)$ were addressed in [77] and it turns out that one needs n to be a divisor of 24 in order to have the n-root of $j(q)$ to have integer coefficients.

This above observation placed a firm relationship between $\mathbb{M}$ and E_8. There is a third, from around the same time, and this will turn out to be the one on which we will focus [55]. Of the 194 conjugacy classes of $\mathbb{M}$, there are 2 which are involutions, i.e., elements therein square to the identity. These are denoted [72] as $2A$ and $2B$ — in general, the conjugacy classes in $\mathbb{M}$ are usually denoted by an integer, signifying the order of the elements, followed by a capital letter which is just an index for classes of the same order. The class $2A$ has $2^4 \cdot 3^7 \cdot 5^3 \cdot 7^4 \cdot 11 \cdot 13^2 \cdot 29 \cdot 41 \cdot 59 \cdot 71 \sim 10^{20}$ elements while $2B$ exceeds it by some 7 orders of magnitude.

If we were to multiply any two elements of $2A$, the resulting element can be in one of only 9 conjugacy classes, viz.,

$$g_1 \cdot g_2 = g_3 , \quad g_1, g_2 \in 2A \rightsquigarrow$$
$$g_3 \in \{1A, 2A, 3A, 4A, 5A, 6A, 4B, 2B, 3C\} . \tag{5.14}$$

Glancing back at the affine Dynkin diagram for $\widehat{E_8}$, we see that they are precisely the (dual Coxeter) labels of the 9 nodes. That is, we have

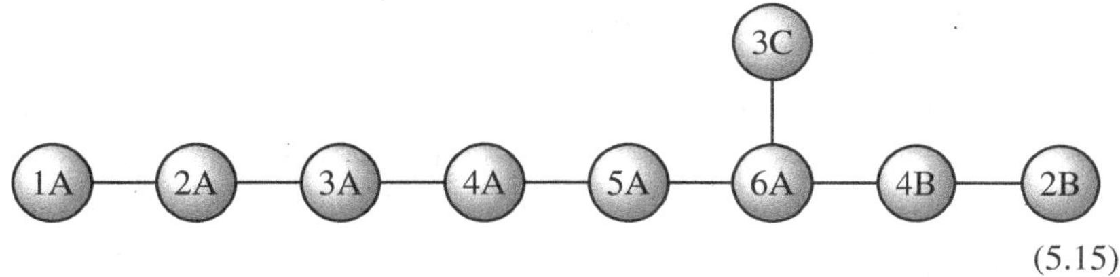

$$\tag{5.15}$$

The edges, i.e., the meaning of adjacency in analogy with the classic McKay Correspondence in (4.12), however, have no clear interpretation in this correspondence which still awaits clarification [78–81]. One difficulty is that the Hauptmoduln for the different nnodes (classes) here are of different modular levels. A recent work [80] recasts this observation solely in terms of the properties of $PSL(2, \mathbb{R})$.

The Baby and $\widehat{E_7}$

An important subgroup of the Monster is the affectionately named Baby Monster, $\mathbb{B}$, of order $2^{41} \cdot 3^{13} \cdot 5^6 \cdot 7^2 \cdot 11 \cdot 13 \cdot 17 \cdot 19 \cdot 23 \cdot 31 \cdot 47$. It is the next largest in the sporadic family. Its double cover, $2.\mathbb{B}$, is the centraliser of an involution in class $2A$ in $\mathbb{M}$.

For reference, we note the centralisers of the relevant conjugacy classes in the Monster and redraw the diagram (and also without the mysterious adjacency) of (5.15):

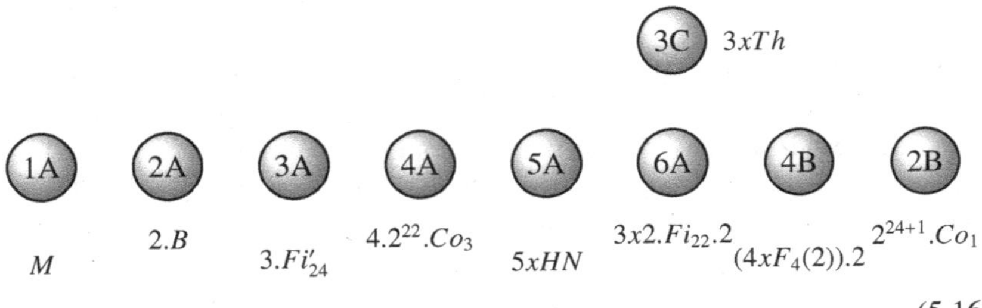

$$(5.16)$$

The observation in (5.15) was generalised by [82] to relate $\mathbb{B}$ to E_7. Here, we have the product of two involution classes of the Baby falling into 8 classes whose orders coincide with the dual Coxeter numbers of affine E_7:

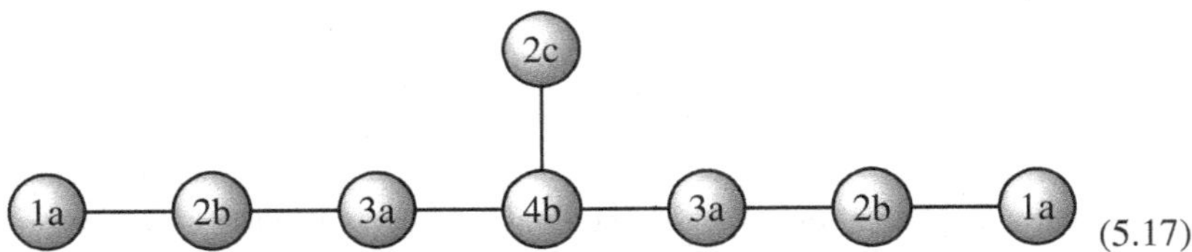

$$(5.17)$$

In the above the class names are those of $\mathbb{B}$ so that in the double cover $2.\mathbb{B}$ some of these split into different conjugacy classes. In particular, whereas we have $(1a, 2b, 3a)$ on one branch in the diagram, the $(1a, 2b, 3a)$ on the other branch are these classes in $\mathbb{B}$ multiplied by the centre $\mathbb{Z}/2\mathbb{Z}$ element when lifting to $2.\mathbb{B}$. In other words, one could read the above diagram modulo the $\mathbb{Z}/2\mathbb{Z}$ centre of $2.\mathbb{B}$. Furthermore, one could *fold* the diagram according to the $\mathbb{Z}_2$ symmetry and thereby associate $\mathbb{B}$ to the Dynkin diagram of $\widehat{F_4}$.

Fischer's Fi'_{24} and $\widehat{E_6}$

The next important subgroup of $\mathbb{M}$ is the largest of the simple Fischer groups, Fi'_{24} (sometimes denoted as F_{3+}), of order $2^{21} \cdot 3^{16} \cdot 5^2 \cdot 7^3 \cdot 11 \cdot 13 \cdot 17 \cdot 23 \cdot 29$. In fact, Fi'_{24} is the next largest one in the sporadic family. Its triple cover $3.Fi'_{24}$ is the centraliser of class $3A$ in $\mathbb{M}$.

Here, the analogue of (5.15) was again generalised by [82] for Fi_{24}, the double cover of Fi'_{24}. In particular, the involution classes multiply to only 7 classes which correspond to the affine $\widehat{E_6}$ labels:

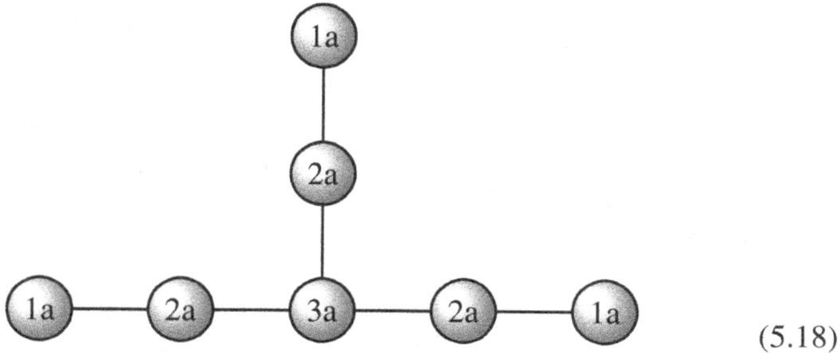

$$(5.18)$$

As above, the class names are those of Fi'_{24} and could split up in the triple cover. In particular while on one branch we have classes $(1a, 2a)$ and those on the other two branches are these multiplied by the $\mathbb{Z}/3\mathbb{Z}$ centre of $3.Fi'_{24}$. Likewise, one could fold according to the $\mathbb{Z}_3$ symmetry reminiscent of triality and associate Fi'_{24} itself to the Dynkin diagram of $\widehat{G_2}$.

5.1.3 Mathieu Moonshine, Umbral Moonshine and the Niemeier Lattices

There have been other recent results that intriguingly connect finite group theory with (mock) modular form theory, number theory and physics. These will temporarily lead us away from ADE only to come back with a twist at the end!

Mathieu Moonshine

In the 2010s, people noticed a connection between Mathieu groups, the geometry of K3 surfaces (see also Section 5.8) and physics. This moonshine phenomenon came from a physical model of a K3 σ-model conformal field theory with an action of the $N = (4, 4)$ superconformal algebra, which arises from a hyper-Kähler structure. The modular object in question is called the elliptic genus, and when decomposing this elliptic genus in terms of the characters of the $N = (4, 4)$ superconformal algebra, akin to the Monstrous Moonshine observation, the multiplicities turn out to be simple combinations of representations of the Mathieu group M_{24} [83]. The $N = (4, 4)$ multiplicities are modular objects: a generalisation of modular forms called mock modular forms [84, 85]. This led Miranda Cheng and others to the Mathieu Moonshine conjecture that characters of M_{24} group elements should likewise be mock modular forms, with the physics giving rise to an infinite-dimensional graded representation like for the Monstrous Moonshine module in Equation (5.8) and an analogue of the McKay–Thompson series (5.10). A wealth of work followed exploring this relationship between K3 geometry, mock modular forms and M_{24} [86–95]. The elliptic genus for K3 is governed by an enumeration function (e.g., for BPS states in a K3 non-linear σ-model) given by a weight 1/2 mock modular form with Fourier expansion

$$2q^{-1/8}(-1 + 45q + 231q^2 + 770q^3 + 2277q^4 + 5796q^5 \dots),$$

where 45, 231, 770, etc. are exactly dimensions of irreducible representations of the sporadic Mathieu group M_{24}.

Umbral Moonshine and the Niemeier Lattices

Cheng and others then noticed that the Mathieu Moonshine phenomenon might be an example of a more general phenomenon termed Umbral Moonshine, which involved Niemeier lattice root systems. The Niemeier lattices are the 24 positive definite even unimodular lattices in dimension 24 [68, 96]. The Leech lattice Λ_{Leech} is the only one of them that is not associated to a root lattice – in fact, the other Niemeier lattices correspond to deep holes inside the Leech lattice.

Each other Niemeier lattice can be constructed from its associated root lattice in a certain "gluing" construction. The root lattices all have rank 24 and are made of ADE building blocks such that all components have the same Coxeter number. So now we are back to ADE! The list of Niemeier root systems is thus

$$A_1^{24}, A_2^{12}, A_3^{8}, A_4^{6}, A_6^{4}, A_8^{3}, A_{12}^{2}, A_{24}, D_4^{6}, D_6^{4}, D_8^{3}, D_{12}^{2}, D_{24}, E_6^{4}, E_8^{3},$$

$$A_5^{4}D_4, A_7^{2}D_5^{2}, A_9^{2}D_6, A_{15}D_9, A_{17}E_7, D_{10}E_7^{2}, D_{16}E_8, A_{11}D_7E_6.$$

The Leech lattice has an automorphism group that is a double cover of the finite simple Conway group Con_1, whilst the A_1^{24} and A_2^{12} lattices are acted on by the Mathieu groups M_{24} and M_{12}, respectively. So it is plausible that the Niemeier lattices could provide a unified link between the representation theory of finite simple groups and mock modular forms.

The idea behind Umbral Moonshine is the following [97–103]. For each Niemeier root system one can define an umbral group, which is the quotient of the automorphism group of the Niemeier lattice by the subgroup of reflections (they are the stabilisers of the deep holes in the Leech lattice). Each umbral group is conjectured to have an infinite-dimensional graded representation (or "module", akin to the Monster module) such that characters (essentially the McKay–Thompson series) are given by vector-valued mock modular forms that satisfy minimality properties akin to the Hauptmoduln condition in Monstrous Moonshine (genus zero). The root systems allow one to construct vector-valued theta series (so-called shadows, hence "umbral") which – because of the minimality properties – uniquely determine the mock modular forms. The Umbral Moonshine conjecture was proven in [104]. The special case of the A_1^{24} lattice yields Mathieu Moonshine. However, the other Niemeier cases do not currently have a geometric or physical interpretation, though there is work relating Umbral Moonshine to du Val singularities on K3 surfaces (see also Section 5.8) via relating the ADE building blocks in the Niemeier lattices to the ADE-type du Val singularities that a K3 surface can have in K3 σ-models [105].

Whilst not an ADE correspondence as such, these connections very much carry on in the vein of binary polyhedral groups, ADE root systems, finite groups, etc. ADE-ology is an exciting field with many interesting connections yet to be discovered. We sincerely hope that this book will help stimulate creative and collaborative new mathematics shedding light on this web of connections. We will now move on to similar advanced topics within the ADE web.

5.2 The Triangle Property

In this section, we will discuss another area of mathematics with strong links to our theme. Indeed, we will see a problem whose solution fits the ADE paradigm of two infinite families and three exceptions; and, moreover, this solution gives rise to a second classification of the ADE root systems, quite different from the approach using simple roots.

A significant figure in this discussion is Jaap Seidel, whom the first author (PJC) regards as a significant mentor and friend. It may be appropriate to tell a personal story here. PJC was at a meeting at the Mathematisches Forschungsinstitut Oberwolfach, in Germany, along with Seidel, the coding theorist Jessie MacWilliams, and many others. The schedule allowed free time between lunch and tea, so participants could discuss mathematics or walk in the forest or both. One day, PJC and Seidel were walking along a track, deep in discussion, when they passed a forester at work. A little later, MacWilliams came along the same track. The forester pointed down the track and said to her, "Your husband and your son went that way." Seidel was delighted by this story.

The story begins with the work of Jacques Tits on geometries associated with the classical groups. These groups act on finite-dimensional vector spaces, preserving non-degenerate forms of various types (alternating bilinear forms, Hermitian forms, or quadratic forms). The geometry associated with such a form, called a *polar space*, consists of all the totally isotropic subspaces for the form (those on which it is identically zero). In a major work, Tits [106] gave an axiomatic description of polar spaces whose rank is at least 3; apart from rank 3, these are just the geometries of the classical groups. This important theorem found many applications; in particular, it aided in recognising the classical groups in various parts of the Classification of Finite Simple Groups.

Independently, Ernie Shult in 1972 gave a characterisation of a class of graphs associated with symplectic and orthogonal geometries over the 2-element field [107]. His arguments were extended by Seidel, who weakened the hypothesis. All these involved recognising the geometries directly. But a huge extension was found by Buekenhout and Shult [108]; their theorem axiomatises the point-line geometries of arbitrary polar spaces, by showing that two very simple axioms imply that Tits' axioms for the entire polar space are satisfied.

Our story will concentrate on the work of Shult and Seidel.

5.2.1 The Strong Triangle Property

A finite graph G is said to have the *strong triangle property* if the following is true: every edge $\{x, y\}$ is contained in a triangle $\{x, y, z\}$ with the property that any further vertex is joined to exactly one of x, y, z. Now the theorem of Shult and Seidel is the following:

THEOREM 5.2 *Let G be a graph with the strong triangle property. Then one of the following holds:*

 (i) *G is a null graph (a graph with no edges);*

 (ii) *G consists of a number of triangles s haring a single common vertex, with no other edges;*

 (iii) *G is one of three exceptional graphs, on 9, 15 or 27 vertices.*

The appearance of the ADE paradigm in this theorem is clear. The graphs in the second case are sometimes called *friendship graphs* because of their occurrence in the Friendship Theorem of Erdős, Rényi and Sós [109]: in a finite society in which any two people have exactly one common friend, there is someone who is everyone else's friend. The shape of such graphs, a windmill, is shown below.

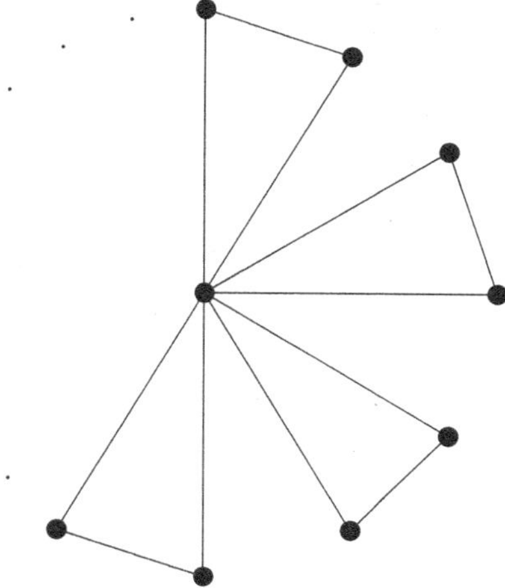

We give a sketch of the proof. Let G be such a graph; we may assume that G is not a null graph. It is easy to see that, if G has a vertex which is joined to all other vertices, then it is as described in the second case of the theorem; so we may assume that this is not the case. Now any two non-adjacent vertices in G have the same valency. For take a vertex v; its neighbourhood consists of a number k of triangles with common vertex v. If w is not joined to v, then w is joined to one point in each of these triangles; and these k neighbours of w lie in k distinct triangles, so w has valency at least $2k$. Reversing the argument shows equality. Now it can be shown that, assuming the second case of the theorem does not occur, all vertices have the same valency; thus G is a *strongly regular* graph with parameters $(n, 2k, 1, k)$. (This means that it is regular with valency $2k$, and two vertices have 1 or k common neighbours according as they are joined or not.) So the adjacency matrix A satisfies $A^2 = 2kI + A + k(J - I - A)$, where J is the all-1 matrix. From this, the eigenvalues and multiplicities can be calculated. The fact that the multiplicities must be positive integers shows that $k = 2, 3, 5$ or 11. An *ad hoc* argument excludes $k = 11$, and the remaining cases give the numbers of vertices given.

For more on strongly regular graphs we refer to [110].

There is a unique graph in each case of part (iii). These can be described as follows:

(i) For $k = 2$, the graph is the 3×3 grid graph: the vertices form a 3×3 array, with two vertices joined if they lie in the same row or column. The distinguished triangles are the rows and columns.

(ii) For $k = 3$, the vertices are the 2-element subsets of a 6-element set A, and two vertices are joined if the subsets are disjoint. The distinguished triangles are the triples whose union is A.

(iii) For $k = 5$, the graph has 27 vertices, which occur in classical algebraic geometry as the 27 lines on a general cubic surface; two lines are adjacent if they intersect. Schläfli gave a combinatorial construction now called *Schläfli's double-six*, which goes as follows. Take a set A of size 6. The vertices are the 2-element subsets of A together with two copies of A, which we denote by A_1 and A_2. The graph on the 15 pairs is the same as in the previous case. A pair $\{a, b\}$ is joined to the points $a_1, b_1 \in A_1$ and $a_2, b_2 \in A_2$ corresponding to a and b.

They occur in many different contexts. For example, the points and triangles in each graph form a *generalised quadrangle*; the proof shows that these are the only generalised quadrangles which have three points on every line.

5.2.2 From the Strong Triangle Property to Root Systems

Now we describe briefly the construction of the ADE root systems from graphs with the strong triangle property.

We take a connected root system Φ having two properties:

- all roots have the same length;
- the crystallographic condition holds, so that the angle between two roots is $0°, 60°, 90°, 120°$ or $180°$.

We choose the lengths of the vectors to be $\sqrt{2}$, so that the inner product of two roots is $2, 1, 0, -1$ or -2.

A *star* consists of six vectors in a plane with angles $0°, 60°, 120°$ or $180°$. Call these vectors a, b, c and their negatives, where $a + b + c = 0$. Using a star, we can decompose the remaining roots into four classes A, B, C, D, according to their inner products with a, b, c:

- $(0, 1, -1)$ or $(0, -1, 1)$;
- $(-1, 0, 1)$ or $(1, 0, -1)$;
- $(1, -1, 0)$ or $(-1, 1, 0)$;
- $(0, 0, 0)$.

Now we show that A, B and C are isomorphic with respect to inner products, and $\{a, b, c\} \cup A$ determines the entire root system. Now take half the vertices in A, those with inner product 1 with b, say, and construct a graph on this set by joining two roots if they are orthogonal. This graph has the strong triangle property, and so is null, a windmill, or one of the three exceptional graphs; and accordingly, the entire root system is of type A, D or E.

The classification can be extended to all root systems satisfying the crystallographic condition. This is based on two observations:

- In any root system, the roots of fixed length form a root system. This is because the reflections in the hyperplanes orthogonal to the roots are length-preserving, and so map roots of any given length to themselves.
- If α and β are independent roots in a crystallographic root system, then both $2\langle\alpha|\beta\rangle/\langle\alpha|\alpha\rangle$ and $2\langle\alpha|\beta\rangle/\langle\beta|\beta\rangle$ are integers, so $s = 4\langle\alpha|\beta\rangle^2/\langle\alpha|\alpha\rangle\langle\beta|\beta\rangle$ is an integer, which is at most 3 since $\langle\alpha|\beta\rangle^2 < \langle\alpha|\alpha\rangle\langle\beta|\beta\rangle$. Thus this number, if not zero, is 1, 2 or 3. So roots of different lengths have lengths in the ratio $\sqrt{2} : 1$ or $\sqrt{3} : 1$.

Thus a crystallographic root system which is not of ADE type is made up of two root systems which are direct sums of ADE root systems, one scaled by a factor $\sqrt{2}$ or $\sqrt{3}$. Further analysis gives only the following possibilities:

- Ratio $\sqrt{2}$, long roots form A_n, short roots form an orthonormal basis (a direct sum of root systems A_1): this is type B_n.
- Ratio $\sqrt{2}$, long roots form an orthonormal basis, short roots form A_n: this is type C_n.
- Ratio $\sqrt{2}$, long and short roots both form D_4: this is F_4.
- Ratio $\sqrt{3}$, long and short roots both form A_2: this is G_2.

5.2.3 The Triangle Property

Now we come to the result of Shult and Seidel, which contains a big surprise. These authors define a triangle property of a graph, a weakening of the strong triangle property already discussed. It turns out that the three exceptional graphs with the strong triangle property are each the first member of an infinite family of graphs with the triangle property.

This involves looking at particular polar spaces over the field $\mathbb{F}_2$ with two elements 0 and 1. A *quadratic function* or *quadratic form* on a vector space V over $\mathbb{F}_2$ is a function $Q\colon V \to \mathbb{F}_2$ satisfying two conditions:

- $Q(0) = 0$;
- the function $B\colon V \times V \to \mathbb{F}_2$ defined by

$$B(x, y) = Q(x + y) - Q(x) - Q(y)$$

is bilinear, i.e., linear in each variable.

The form is *non-singular* if the only vector $v \in V$ which satisfies $Q(v) = 0$ and $B(v, w) = 0$ for all $w \in V$ is the zero vector.

There is a classification of non-singular quadratic forms over $\mathbb{F}_2$, which we now outline. A subspace W of V is *anisotropic* if $Q(w) \neq 0$ for all $w \in W$. A *hyperbolic plane* is a 2-dimensional subspace spanned by two vectors v, w with $Q(v) = Q(w) = 0$ and $B(v, w) = 1$. (The name comes from the fact that, for any scalars x, y, we have $Q(xv + yw) = xy$.)

Now given a space V with a non-singular quadratic form, we can express V as the orthogonal direct sum of an anisotropic space and a number r of hyperbolic planes; the number r and the isomorphism type of the anisotropic space are the same for any such decomposition. The number r is called the *Witt index*. Now an anisotropic space over $\mathbb{F}_2$ has dimension at most 2. (To prove this, we simply have to show that the function on a 3-dimensional space which is 0 at the origin and 1 everywhere else is not a quadratic form.) So any non-singular quadratic space is isomorphic to $V_{r,i}$ where $r \geq 0$ and $i \in \{0, 1, 2\}$; here r is the Witt index and i the dimension of the anisotropic space. It can be shown that the Witt index is equal to the maximum dimension of a space on which the form is identically 0.

The values $= 0, 1, 2$ are related to the trinity we met briefly in Section 3.1. The three types of quadric are called hyperbolic, parabolic and elliptic, respectively.

Now there is a graph associated with such a space V, with vertex set $\{v \in V : v \neq 0, Q(v) = 0\}$, in which two vertices v and w are joined if and only if $B(v, w) = 0$. We will call this graph $\mathcal{G}_{r,i}$, where the indices have the same meaning as above. Note that, if $r = 0$, then the space is anisotropic, and the graph has no vertices. If $r = 1$, then the graph consists of 2, 3 or 5 isolated vertices. For $r = 2$, the three graphs turn out to be exactly those in the strong triangle property theorem. To illustrate, here is the graph $\mathcal{G}_{2,0}$, with the vertices labelled by vectors in $\mathbb{F}^4$. (The quadratic form is $x_1 x_2 + x_3 x_4$, and the associated bilinear form is $x_1 y_2 + x_2 y_1 + x_3 y_4 + x_4 y_3$. Check that the three points in a row or column are singular points which are orthogonal with respect to the bilinear form and sum to zero.)

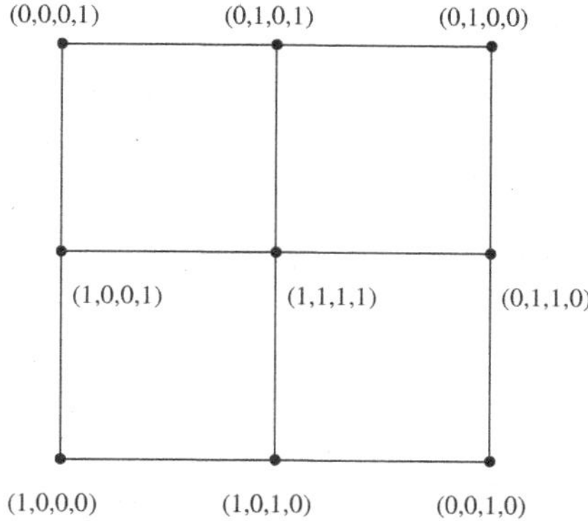

A graph G is said to have the *triangle property* if every edge $\{x, y\}$ is contained in a triangle $\{x, y, z\}$ with the property that any further vertex is joined to one or all of x, y, z. Now we can state the Shult–Seidel theorem:

THEOREM 5.3 *Let G be a finite graph with the triangle property. Then one of the following is true:*

  (i) G *is a null graph (a graph with no edges);*
 (ii) G *has a vertex which is joined to all others;*
(iii) G *is isomorphic to $G_{r,i}$ for some $r \geq 2$ and $i \in \{0, 1, 2\}$.*

So at least in this case of the ADE paradigm, the type E objects turn out not to be isolated exceptions, but the start of three infinite families.

5.3 Graphs with Least Eigenvalue −2 or Greater

In Section 3.3 we saw the ADE and affine patterns arise in graph-theoretic problems: as the graphs whose largest eigenvalue λ_1 is 2 or smaller. What about the "dual" problem: connected graphs whose adjacency matrix has least eigenvalue −2 or greater. In our notation earlier, we need to consider graphs with $\lambda_n \geq -2$. This was a topic of great interest in the 1960s, especially in the work of Alan Hoffman. We here discuss the solution obtained by [35].

First we observe that information about the smallest eigenvalue can tell a lot about a graph, and that geometric methods can be useful.

The *Gram matrix* of a set $\{v_1, \ldots, v_n\}$ of vectors in a real inner product space is the $n \times n$ matrix whose (i, j) entry is $v_i \cdot v_j$.

THEOREM 5.4 *Let A be an $n \times n$ real symmetric matrix which is positive semidefinite of rank r. Then A is the Gram matrix of a set of n vectors in $\mathbb{R}^r$.*

Proof There is an invertible matrix P such that

$$P^\top A P = \begin{pmatrix} I_r & O \\ O & O \end{pmatrix}.$$

Writing

$$P^{-1} = \begin{pmatrix} R \\ S \end{pmatrix}$$

where R is $r \times n$, we see that $A = R^\top R$; so the required vectors are the columns of R. $\qquad\square$

Let G be a non-trivial graph whose adjacency matrix A has least eigenvalue $-\lambda$, with multiplicity $n - r$. Then $A + \lambda I$ is positive semi-definite of rank r; so there exist n vectors $v_1, \ldots, v_n \in \mathbb{R}^r$ such that $v_i \cdot v_i = \lambda$, while for $i \neq j$ we have $v_i \cdot v_j = 1$ if i and j are joined, 0 otherwise. Thus the vectors $v_1, \ldots, v_n$ have squared length λ, and make an angle $\cos^{-1}(1/\lambda)$ if i and j are joined, or $\pi/2$ otherwise.

So now let G be a graph with least eigenvalue -1 (or greater). Then this process gives us a set of unit vectors with inner products 1 (if adjacent) or 0 (otherwise). Since unit vectors with inner product 1 are equal, we see that the graph consists of a disjoint union of complete graphs.

We now turn to the case where the least eigenvalue is -2 (or greater), and give the solution to Hoffman's problem.

Suppose that $\mathcal{G}$ is a connected graph whose adjacency matrix $\mathcal{A}$ has smallest eigenvalue -2 or greater. Then $2I + \mathcal{A}$ is positive semi-definite and so is, by the same reasoning as above, the Gram matrix of a set $\{v_1, \ldots, v_n\}$ of vectors in a real inner product space. Since the diagonal is 2, every vector has length $\sqrt{2}$, and any two vectors lie at an angle of $\cos^{-1}(0) = 90°$ or $\cos^{-1}((\frac{1}{\sqrt{2}})^2) = 60°$, since the off-diagonal elements are 0 and 1.

These angles immediately impose a severe constraint, as we see from

LEMMA 5.5 *Let S be a set of vectors with length $\sqrt{2}$ in $\mathbb{R}^d$, any two making angles $60°$, $90°$, $120°$ or $180°$.*

(i) *The set S is contained in a set $\hat{S}$ with this property, which contains the negative of each of its members and contains all six vectors in the plane spanned by two of its vectors at angle $60°$ or $120°$.*

(ii) *The set $\hat{S}$ just described is a simply-laced root system, i.e., an orthogonal direct sum of root systems of type A, D or E.*

Proof The first part of the lemma is proved by first adjoining the negatives of all the vectors in S, and then adjoining $\pm(v + w)$ for each pair v, w with

$v \cdot w = -1$ (if these vectors are not already included). In this step we have to show that no conflict is created. Suppose that z is another vector in the set; then $v \cdot z$ and $w \cdot z$ belong to $\{-1, 0, +1\}$, so the only problem would be if, say, both these inner products were $+1$. But then $v + w$ and z would be two vectors of length $\sqrt{2}$ with inner product 2; so they are equal. In other words, the set of lines is star-closed: that is, if two lines in the set make angle 60°, then the third line in their plane at 60° to both is also in the set. The second part of the lemma is clear from the definition of a root system. Thus, graphs with least eigenvalue −2 are "contained" in a root system of type ADE. □

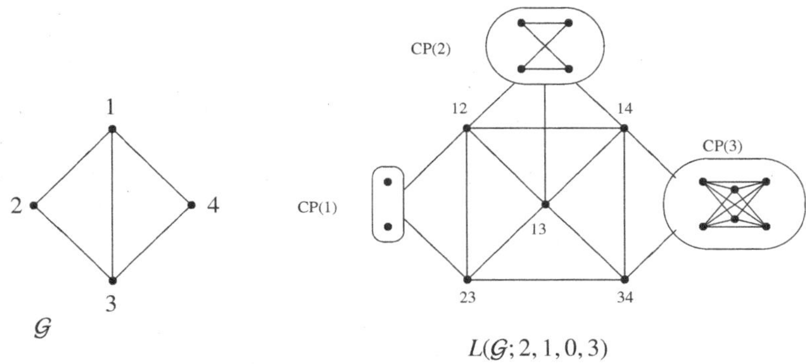

Figure 5.1   Example of a generalised line graph.

Next, we will need the notion of a line graph:

DEFINITION 5.6   Given a graph G with vertex set V and (undirected) edge set E, the *line graph* of G (written $L(G)$) is the graph with vertex set E, in which two edges are adjacent if and only if they share a vertex in V.

As an example, consider the LHS of Figure 5.1. For this graph G, the vertex set is $V = \{1, 2, 3, 4\}$ and the edge set is $E = \{E_{1,2}, E_{1,3}, E_{1,4}, E_{2,3}, E_{3,4}\}$. So the line graph $L(G)$ would have 5 vertices, which we can denote by the pairs $\{12, 13, 14, 23, 34\}$. Then, two nodes have an edge between if they share an index because in the original G they would have shared a node. In particular, the edges are 12–13, 12–14, 12–23, 13–14, 13–23, 13–34, 14–34, 23–34. These 8 edges are drawn in the center of the RHS of the figure.

In general, let $V = \{v_1, \ldots, v_n\}$ be the vertex set of G, we can represent $L(G)$ by a subset of the space $\mathbb{R}^n$ with standard orthonormal basis $\{e_1, \ldots, e_n\}$, where the edge $\{v_i, v_j\}$ is represented by the vector $e_i + e_j$. We see that

$(e_i + e_j) \cdot (e_k + e_l) \in \{0, 1\}$, the value being 1 if and only if the edges $\{v_i, v_j\}$ and $\{v_k, v_l\}$ share a vertex. Thus, the Gram matrix of the set of vectors is $2I + \mathcal{A}(L(\mathcal{G}))$; so we see that the smallest eigenvalue of the adjacency matrix $\mathcal{A}(L(\mathcal{G}))$ is at least -2.

Another graph with smallest eigenvalue -2 is the so-called *cocktail party graph* (see, e.g., [111])

DEFINITION 5.7   A cocktail party graph $CP(m)$ is a connected (undirected) finite graph with $2m$ vertices $v_1, \ldots, v_m, w_1, \ldots, w_m$, such that all pairs are joined except $\{v_i, w_i\}$ for $i = 1, \ldots, m$.

In other words, we draw two rows of nodes, $v_{i=1,\ldots,m}$ and $w_{i=1,\ldots,m}$ and pair each $\{v_i, w_i\}$. Then, we draw all possible edges, except for those which are paired. Think of inviting m couples to a cocktail party and for best mixing everyone is allowed to talk to anyone except their partner. Indeed, $CP(m)$ can be represented in Euclidean space by the vectors $e_0 \pm e_i$ for $i = 1, \ldots, m$. Thus its adjacency matrix also has smallest eigenvalue -2.

In the 1960s, Alan Hoffman (q.v. [112]) combined these two above-mentioned graphs into the notion of a *generalised line graph*.

DEFINITION 5.8   Start with a graph $\mathcal{G}$, with a non-negative integer a_i associated with each vertex v_i. The generalised line graph $L(\mathcal{G}; a_1, \ldots, a_n)$ is built as follows: (1) we take the disjoint union of $L(\mathcal{G})$ with cocktail party graphs $CP(a_1), \ldots, CP(a_n)$, and (2) add edges joining vertices of $CP(a_i)$ to vertices of $L(\mathcal{G})$ corresponding to edges of $\mathcal{G}$ containing the vertex v_i.

Figure 5.1 continues with our example, showing how the cocktail party graphs and the line graph associated to $\mathcal{G}$ combine to give the generalised line graph $L(\mathcal{G}; 2, 1, 0, 3)$. It is an exercise for the reader to find a collection of vectors in Euclidean space representing a generalised line graph. (See Exercise 5.2.)

Hoffman guessed that all "sufficiently large" (in some sense) connected graphs with smallest eigenvalue -2 or greater are generalised line graphs. By "sufficiently large", he meant that the minimum degree of the graph should be sufficiently large. The next theorem shows that Hoffman's guess is correct, in the much stronger sense that "sufficiently large" means "not represented in the root system E_8", so there are only finitely many exceptions.

THEOREM 5.9   *Let $\mathcal{G}$ be a connected graph with smallest eigenvalue -2 or greater. Then either*

- $\mathcal{G}$ *is a generalised line graph; or*
- $\mathcal{G}$ *is represented by a subset of the root system E_8.*

Proof (sketch) The theorem follows easily from our observations. Represent the adjacency matrix of G as the Gram matrix of a set S of vectors, and enlarge S to a root system $\hat{S}$. Since G is connected, $\hat{S}$ is indecomposable, so it is ADE. Now to conclude the proof we note two further facts:

- $A_n \subseteq D_{n+1}$, and $E_6 \subseteq E_7 \subseteq E_8$;
- a graph is represented by a subset of D_n if and only if it is a generalised line graph.

The first part is clear from our explicit representations of these root systems given earlier. Here is an outline proof of the second. Recall that D_n is represented by the vectors $\pm e_i \pm e_j$ for $i \neq j$. Now, if a basis vector occurs with the same sign in all roots in the set, we can change its sign if necessary and assume that it is positive. We call such basis vectors of vertex type; the roots involving two of them correspond to edges of a graph G, and represent the line graph of G. Any other basis vector e_j occurs only in roots $e_i + e_j$ and $e_i - e_j$, where e_i is of vertex type; for fixed i, these roots represent a cocktail party graph. So the whole is a generalised line graph. $\qquad\square$

It is known that a graph represented in the root system of E_8 has at most 36 vertices, and valencies at most 28. For regular graphs, these numbers can be reduced to 28 and 16 respectively. Moreover, all regular graphs in E_8 have been explicitly found: see [113]. The following theorem uses the result of Exercise 5.4.

THEOREM 5.10 *A regular graph with smallest eigenvalue −2 or greater is a line graph, a cocktail party graph, or one of a list of 187 explicitly known exceptions (all these exceptions being represented in the root system E_8).*

A number of earlier results, by Hoffman, Ray-Chaudhuri, Seidel, Chang, and others, are subsumed by these above theorems. We refer to [35, 110, 114] for details. For an illustration, we give a theorem of Shrikhande.

THEOREM 5.11 *Let G be a graph with the same spectrum as the line graph of the complete bipartite graph $K_{n,n}$. Then, if $n \neq 4$, G is isomorphic to $L(K_{n,n}$. If $n = 4$, there is exactly one further graph with this spectrum (up to isomorphism).*

The technique can also be applied to find abelian subgroups generated by root groups. In a finite group of Lie type in characteristic p (or any Chevalley group), the Sylow p-subgroup (or maximal unipotent subgroup) is generated by *root subgroups* U_α, which in the ADE case are isomorphic to the additive group of the field. The commutator $[U_\alpha, U_\beta]$ is generated by the root subgroups

U_γ such that γ is a positive combination of α and β [73, Theorem 5.2.2]. In the ADE case, such γ exists only if $\alpha \cdot \beta = -1$. So the root subgroups U_α for $\alpha \in S$ generate an abelian group if and only if $\alpha \cdot \beta \geq 0$ for all $\alpha, \beta \in S$. Such sets correspond to graphs with least eigenvalue -2 or greater.

As a footnote, we observe that Hoffman subsequently proved a different strengthening of his original conjecture, in the paper [115]:

THEOREM 5.12   *A connected graph whose smallest eigenvalue exceeds $-1 - \sqrt{2}$ and whose minimum degree is sufficiently large is a generalised line graph.*

In other words, the lower bound -2 for the eigenvalues is replaced by the smaller $-1 - \sqrt{2}$. It is not clear how to achieve this with the root system technology.

Another area in which root systems play a role is an extension of a theorem of Whitney [116], who showed that, with a single exception, if connected graphs G_1 and G_2 have isomorphic line graphs, then G_1 and G_2 are themselves isomorphic (and, moreover, any isomorphism from $L(G_1)$ to $L(G_2)$ is induced by an isomorphism from G_1 to G_2). The exception[2] is the pair $(K_3, K_{1,3})$ of graphs (a triangle and a 3-star), which have isomorphic line graphs.

This result has been extended to Hoffman's generalised line graphs in [117]. We outline the argument. Represent the graph in the root system D_n as described earlier. Then vertices can be represented by unit vectors making an angle $45°$ or $90°$ with all the vectors representing the graph. The resulting set is contained in a root system with roots of two lengths, 1 and $\sqrt{2}$. If $n \neq 4$, the only possibility is B_n, and the vertices can be recovered uniquely. The exceptions arise from the exceptional root system F_4, which is obtained from D_4 by adjoining three sets of eight vectors each consisting of an orthonormal basis and the negatives of its vectors. All exceptions can be classified. For example, $CP(3) = L(K_1; 3)$ is isomorphic to $L(K_4)$.

Having thus revisited graphs and ADE connections, we now turn our attention to multigraphs, or quivers, and will in turn see an ADE pattern emerge.

5.4 Quiver Representations

Since we have already delved into finite graphs, it is perhaps easiest to begin with *quivers*. These are simply directed graphs where (1) nodes are assigned

---

[2] Recall the standard notation that K_n is the complete graph on n nodes and $K_{m,n}$ is the complete bi-partite graph on 2 sets of nodes, of sizes m and n respectively, such that all nodes in one set are connected to all nodes in the other set, but no connectivity within the nodes of either.

vector spaces and (2) arrows between them are assigned linear maps between the corresponding vector spaces. They have turned out to be a powerful tool in representation theory by "visualising" complicated morphisms between algebraic objects (see also [118–120]).

Formally, quivers (or the original German, *Köcher*) were introduced by Gabriel [121, 122] because the nodes are reminiscent of holders of arrows. Specifically, we have:

DEFINITION 5.13 A **quiver** is a pair $Q = (Q_0, Q_1)$, where Q_0 is a set of vertices (or nodes) and Q_1 a set of arrows. The arrows link vertices in the sense that each element $\alpha \in Q_1$ has a beginning $s(\alpha)$ and an end $e(\alpha)$ which are vertices, i.e., $\{s(\alpha) \in Q_0\} \xrightarrow{\alpha} \{e(\alpha) \in Q_0\}$.

In graph-theoretical language, Q is a directed multi-graph (allowing multiple arrows between nodes). Note that we do not require that either $|Q_0|$ or $|Q_1|$ be finite.

So far, we have merely been introducing a graph. The key point is to associate with these the aforementioned algebraic objects such as vector spaces and maps:

DEFINITION 5.14 The **representation of a quiver** rep(Q) of Q is the assignment: (1) to each vertex $x \in Q_0$ of Q a vector space V_x, and (2) to each arrow $x \to y$ a linear transformation (morphism) between the corresponding vector spaces $V_x \to V_y$.

Finally, we can create an algebra from a quiver. Given a ground field k (which, we can take to be $\mathbb{C}$) and a quiver Q, a **path algebra** kQ is an algebra which as a vector space over k has its basis prescribed by the paths in Q. An algebra needs a bilinear product; for kQ, it is the concatenation of paths if the end of one is the beginning of the other, and 0 (the zero-vector) otherwise. Note that kQ is a finite-dimensional algebra if Q is finite and acyclic (so that we do not have infinite paths by going around cycles).

Finally, we recall a classic theorem [123] on finite-dimensional algebras A:

THEOREM 5.15 *(Trichotomy theorem of representation type) Over an algebraically closed field k, every finite-dimensional algebra A is one of the 3 types: finite, tame or wild.*

Of course, the theorem requires a definition of the 3 types; these are as follows. An algebra A is of

(i) **finite** representation type if there are only finitely many isomorphism classes of indecomposable A-modules, otherwise it is of infinite type.

(ii) **tame** representation type if it is of infinite type and for any dimension n, there is a finite set of A-$k[X]$-bimodules M_i obeying the following:[3]

- M_i are free as right $k[X]$-modules;

- For some i and some indecomposable $k[X]$-module M all but finitely many indecomposable A-modules of dimension n can be written as $M_i \otimes_{k[X]} M$.

(iii) **wild** representation type if it is of infinite representation type and there is a finitely generated A-$k[X, Y]$-bimodule M which is free as a right $k[X, Y]$-module such that the functor $M \otimes_{k[X,Y]}$ from finite-dimensional $k[X, Y]$-modules to finite-dimensional A-modules preserves indecomposability and isomorphism classes.

These definitions are clearly overwhelmingly technical but details are not relevant for our present discussions. We will only illustrate with a few examples. Consider the algebra $A = \mathbb{C}[x]/\langle x^n \rangle$, the algebra of complex polynomials (under usual polynomial multiplication and addition) in a single variable x such that $x^n = 0$ for some integer $n \geq 1$. That is, A is the algebra of complex polynomials up to degree $n - 1$. Any A-module M is a vector space together with a linear map (matrix) $\phi \curvearrowright M$ such that $\phi^n = 0$. Put ϕ in, say, Jordan canonical form, then the indecomposables correspond to the Jordan blocks, of which there is clearly a finite number.

On the other hand, consider $A = \mathbb{C}[x, y]/\langle x^2, y^2 \rangle$. Then $M = \mathbb{C}^{2n}$ for some integer $n \geq 1$ is an A-module with action by matrices $X := \left(\begin{smallmatrix} 0 & I_n \\ 0 & 0 \end{smallmatrix} \right)$ and $Y := \left(\begin{smallmatrix} 0 & J_n \\ 0 & 0 \end{smallmatrix} \right)$. Here, I_n is the $n \times n$ identity matrix and J_n is an $n \times n$ matrix with some fixed $\lambda \in \mathbb{C}$ on the diagonal and 1s on the upper off-diagonal (i.e., $(J_n)_{i,i} = \lambda$ and $(J_n)_{i,i+1} = 1$ with 0 everywhere else). One easily checks that $X^2 = Y^2 = XY - YX = 0$ so that M is indeed an A-module. Moreover, M is indecomposable. Clearly, for (the infinitely many possible) different values of n and λ, M are non-isomorphic. Thus, here A is of infinite representation type.

The point is that the definitions of the trichotomy seem completely unrelated to ADE-ology, but a pair of remarkable theorems[4] brings us back to our familiar theme:

---

[3] Therefore for the polynomial ring $k[X]$, the indeterminate X furnishes a one-parameter family and the indecomposable $k[X]$-modules are classified by powers of irreducible polynomials over k. If the M_i may be chosen independently of n, then we say A is of *domestic* representation type.

[4] Here as usual, each edge of the Dynkin diagrams, as a quiver, can be taken as a bi-directional arrow between associated nodes.

THEOREM 5.16 (Gabriel [121, 122]) *Given a finite quiver Q, its path algebra kQ is of* **finite** *representation type if and only if it is a disjoint union of Dynkin graphs of type A_n, D_n and $E_6/E_7/E_8$, i.e., the ordinary simply-laced ADE Coxeter–Dynkin diagrams.*

In parallel, we have

THEOREM 5.17 (Nazarova [124–127]) *Let Q be a connected quiver without oriented cycles, then kQ is of* **tame** *(in fact domestic) representation type if and only if Q is one of the graphs of type $\widehat{A}_n$, $\widehat{D}_n$ and $\widehat{E}_6/\widehat{E}_7/\widehat{E}_8$, i.e., the affine ADE Coxeter–Dynkin diagrams.*

Therefore, perhaps as surprising as the analogous theorems 3.9 of J. Smith in 3.3 wherein a simple bound on eigenvalues of undirected graphs led to ADE and affine ADE, representation type on directed graphs (quivers) likewise leads to our protagonists.

Furthermore, due to the underlying representation theory of the McKay Correspondence discussed in 4.5, the graphs constructed from the finite discrete subgroups of SU(2) – or for that matter any analogous graphs constructed from the irreducible representations of a finite group – have come to be called **McKay quivers**.

As with all beautiful objects in mathematics, McKay quivers have found their place in theoretical physics, especially in string theory [128, 129]. We will see this again in Section 5.8. The role of the trichotomy theorem in the context of quantum field theories discussed in [130].

5.5 Cluster Algebras

While we are on the subject of representation theory, one cannot resist but to give an account of "cluster algebras" which have recently made a significant impact across mathematics. The reader is referred to the original works of Fomin and Zelevinsky [131, 132] as well as recent reviews and other work [133–138] for a comprehensive treatment of the material.

We will introduce cluster algebras in their most elementary guise,[5] which will entice the reader as to how something so fundamental (much like Theorem 3.9) could be discovered so late.

---

[5] Cluster algebras originally came up in the theories of Poisson algebras and totally positive matrices, and have connections with algebras of finite representation type, with critical points of smooth functions, as well as algebraic geometry. Indeed, when Fomin and Zelevinsky invented cluster algebras, they at first did not know that the finite-dimensional ones fitted the ADE classification; this was discovered later.

Consider the sequence given by the (non-linear) recurrence

$$y_{n+2} = \frac{1 + y_{n+1}}{y_n}, \quad n \in \mathbb{Z}_{>0}, \tag{5.19}$$

where y_n are formal variables. It is easy to see that it returns to its initial value after five steps. For example, if the first two terms are $y_0 = 1$, $y_1 = 1$, then the sequence runs as $1, 1, 2, 3, 2, 1, 1, \ldots$. More analytically, we can see that the sequence proceeds as: $y_1, y_2, y_3 = \frac{1+y_2}{y_1}, y_4 = \frac{1+y_3}{y_2} = \frac{1+y_1+y_2}{y_1 y_2}, y_5 = \frac{1+y_4}{y_3} = \frac{1+y_1}{y_2},$ and $y_6 = \frac{1+y_5}{y_4} = y_1, \ldots$. In addition to the period 5 property, we also see a "Laurent phenomenon", that each term, despite being ratios of polynomials, will always simplify so that the denominator is a single monomial, whereby making each term a Laurent polynomial. In the numerical example, the Laurent phenomenon exhibits itself as the sequence being all integers, despite having a denominator in the recursion. Note that these two properties are highly sensitive to the nature of the recurrence; changing the 1 to a 2 in (5.19) would completely ruin both.

Now, take the related recurrence

$$y_{n+3} = \frac{1 + y_{n+1} y_{n+2}}{y_n}, n \in \mathbb{Z}_{>0}. \tag{5.20}$$

We can check that the Laurent phenomenon still holds, but the periodicity is gone. For instance, starting $1, 1, 1$, the sequence runs as $1, 1, 1, 2, 3, 7, 11, 26, 41, 97, 153, 362, 571, \ldots$, and grows forever.[6]

What is going on here? The insight of [131, 132] is that both above examples are associated with an underlying algebraic structure called a **cluster algebra**:

DEFINITION 5.18   Let $F := k(y_1, \ldots, y_n)$ be the field of rational functions in variables y_i over a ground field k, which we can take to be $\mathbb{Q}$, for instance. Then we define the following:

- A *cluster* of rank m is a set $\{x_i\}_{i=1,\ldots,m}$ of elements of F.

- A *seed* is a cluster together with an $m \times m$ *exchange matrix* $B = (b_{ij})$, typically taken to be antisymmetric.

Usually, algebras are given as generators with some relations, which generate the whole algebra. Cluster algebras are unusual in that we are given an initial seed along with a prescription for how to generate the other generators.

---

[6] This sequence has many interesting interpretations, including the denominators of the continued fraction convergents to $\sqrt{3}$ (see OEIS sequence A140827.)

We can generate another cluster from a cluster we already have by "mutating in the k-direction". This just replaces x_k in the original cluster by x'_k, where the latter is given by the following exchange relation:

DEFINITION 5.19 An *exchange relation* between x_k and x'_k is given via the B-matrix entries as

$$x_k x'_k = \prod_{\substack{b_{ik}>0 \\ 1\leq i\leq m}} x_i^{b_{ik}} + \prod_{\substack{b_{ik}<0 \\ 1\leq i\leq m}} x_i^{-b_{ik}}.$$

It also mutates the corresponding exchange matrix in the following way. Let $B = (b_{ij})$ and $B' = (b'_{ij})$ be integer matrices. We say that B' is obtained from B via a *matrix mutation*[7] in the direction of k, if

$$b'_{ij} = \begin{cases} -b_{ij} & \text{if } k \in \{i, j\}, \\ b_{ij} + |b_{ik}|b_{kj} & \text{if } k \notin \{i, j\} \text{ and } b_{ik}b_{kj} > 0, \\ b_{ij} & \text{otherwise.} \end{cases}$$

We write $B' = \mu_k(B)$; it is easy to show that such μ_k are involutions.

This prescription leads to a set of matrices related via mutation. In general, mutations of clusters and matrices (i.e., seeds) can generate infinite structures. For cluster algebras there are therefore two types of finiteness: that of clusters and that of matrices. If there are finitely many clusters (and thus seeds), then the cluster algebra is said to be of *finite type*. There is then automatically a finite number of matrices. However, even if there are infinitely many clusters, it may still be the case that there are only finitely many matrices. These infinite cluster algebras are called of *finite mutation type*.

One could think of mutation graphically, and we are back to the situation of quivers. The exchange matrix B can be thought of as the anti-symmetrised adjacency matrix of a finite directed graph (quiver). That is, we take $B_{ij} = a_{ij} - a_{ji}$ for the ordinary adjacency matrix a_{ij} which we have encountered throughout the book. The anti-symmetry of B is so that the graph is without loops and 2-cycles.[8] Thus, the quiver associated to (5.19) is the A_2 Dynkin diagram,[9] while that associated to (5.20) is the oriented triangle. The mutation rule on B then

---

[7] A complete surprise is that in the same year as [131] and completely unaware of each other's work, physicists stumbled on these mutation conditions when studying Seiberg Duality for $\mathcal{N} = 1$ supersymmetric gauge theories [139]. It was only many years later at Oberwolfach when the physicists met Fomin that it was realised everyone was working on the same structure. This is yet another of the many instances why string theory is believed by pure mathematicians to be "on the right track".

[8] In the physics [139, 140], however, we do allow both.

[9] But note that the period 5 in the first case seems to have nothing to do with the root system, or reflection group, or anything else, traditionally associated with A_2. There are groups here too, which are not the Coxeter groups of the appropriate types. For the second example, we get the group of rotations of the icosahedron.

translates to the following graph move: (1) pick a node; (2) reverse all arrows incident on the node; (3) complete the outgoing and incoming arrows to that node by adding an arrow so as to form a loop; (4) erase pairs of antiparallel edges (loops) that may have been created in the process.

In the spirit of our ADE theme – but still highly surprising – the cluster algebras of finite type follow an ADE pattern. We will therefore finish our discussion of cluster algebras here with the highlight theorem and defer the interested reader to explore further details in the literature:

THEOREM 5.20 (Finite type classification) *A cluster algebra $\mathcal{A}$ is of finite type if and only if the exchange matrix B at some seed of $\mathcal{A}$ is the adjacency matrix of an ADE diagram, i.e., $B - 2I$ is the Cartan matrix of an ADE Dynkin diagram.*

The finite type classification therefore follows the same ADE pattern as the classification of simply-laced Lie algebras and the corresponding crystallographic root systems and Coxeter groups.

5.6 von Neumann Algebras and Subfactors

In another central idea in representation theory our diagrams take the stage. We first recall that given a Banach $*$-algebra M with a unit (i.e., an associative algebra over $\mathbb{C}$ which is also a complete metric space with metric induced by a norm $\|\cdot\|$, where the unit has norm 1),

DEFINITION 5.21 A finite **von Neumann algebra** is such an M but also with a trace $\mathrm{Tr}\colon M \longrightarrow \mathbb{C}$, satisfying, for any $a, b \in M$,

  (i) $\|a^*a\| = \|a\|^2$,
 (ii) $\mathrm{Tr}(ab) = \mathrm{Tr}(ba)$,
(iii) $\mathrm{Tr}(1) = 1$,
(iv) $\mathrm{Tr}(a * a) > 0 \qquad \forall a \neq 0$,
 (v) the metric is defined by $\langle a, b \rangle = \mathrm{Tr}(b^*a)$ and is complete.

If M has only a trivial centre, i.e., consisting of only scalar operators, then M is called a **factor**.

Remark: Interestingly, von Neumann was originally inspired by quantum mechanics in order to axiomatically establish his algebra. Perhaps the most important finite von Neumann algebra is when it is infinite-dimensional and with one-dimensional center;[10] this is called a **Type II$_1$ factor**. A **subfactor**

---

[10] In physics, this algebra finds its place in observables for anti-de-Sitter space [141].

N of a II_1 factor M is a sub-$*$-algebra containing the identity and which is itself II_1. Finally, we are interested in ([142], see [143] for a comprehensive account):

DEFINITION 5.22 A subfactor $N \subset M$ is of **finite index** if M is a finitely generated projective left N-module under (left) multiplication. The index $[M : N]$ is then the trace of some idempotent in a matrix algebra defining M as a (left) N-module.

Remarkably, classification of von Neumann subfactors of II_1 for index less than 4 reduces to that of graphs with eigenvalues less than 2. Hence,

THEOREM 5.23 *von Neumann subfactors of index $[M : N] < 4$ are in one-to-one correspondence with the (ordinary) ADE Dynkin diagrams.*

Index greater than or equal to 4 becomes more subtle and the reader is referred to [143].

As a parting remark, before we venture away from the land of discrete mathematics, there is a notion of flat curvature for (locally finite) graphs, generalising that of Ricci-flatness for differential manifolds [144, 145]. One notices [32] that for girth greater than or equal to 5, such flat graphs fall into *somewhat* of an ADE pattern, while for smaller girth, they are infinite in number.

5.7 Catastrophes: Arnold Singularities

Initiated by Thom in the theory of dynamical stability in the 1960s [146] and popularised by Zeeman [147], catastrophe theory has found applications to vast areas, from dynamical systems to differential geometry to applied mathematics, and leads to yet another ADE pattern. The classical introductions are given in [148–150].

The protagonist here is something clearly ubiquitous: the multi-variate function $f \colon \mathbb{R}^n \to \mathbb{R}$. In particular, we are interested in its critical points. First, we recall some rudiments of Morse theory.

DEFINITION 5.24 A *singular point* $p \in \mathbb{R}^n$ of a function $f \colon \mathbb{R}^n \to \mathbb{R}$ is one where the gradient vanishes:

$$df|_p = \left(\frac{\partial f}{\partial x_1}, \ldots, \frac{\partial f}{\partial x_n} \right)\bigg|_p = 0 \,.$$

Moreover, f is non-degenerate, or **Morse**, at p, if the Hessian matrix

$$d^2 f\big|_p = \left(\frac{\partial^2 f}{\partial x_i \partial x_j}\right)\bigg|_p$$

of second derivatives is non-singular there.

Then, we have that (shifting p to be the origin):

THEOREM 5.25 *Let 0 be a non-degenerate singular point of f, then there exist normal coordinates in the neighbourhood of 0 in $\mathbb{R}^n$ such that*

$$f(x) = f(0) - \sum_{i=1}^{k} x_i^2 + \sum_{i=k+1}^{n} x_i^2 \,.$$

Here, k, the index of the quadratic form, is the number of negative eigenvalues of $d^2 f\big|_{x=0}$. Furthermore, for and only for a Morse function, the singularity is **stable**, *in the sense that a small perturbation of f has the same index k.*

From now on, we will (by appropriate shift) take the singular point to be 0. Let us illustrate with a simple example. Consider $f(x) = x^3/3$. At $x = 0$ (which is a singular point since $df(0) = 0$), it is clearly *not* Morse since $d^2 f(0) = 0$. Take a small perturbation $g(x) = x^3/3 - \epsilon x$, where $dg = x^2 - \epsilon$. Thus, if $\epsilon < 0$, there are no singular points at all. If $\epsilon > 0$, there are two singular points $\pm\sqrt{\epsilon}$, both of which are non-degenerate. Thus, the non-Morse f is not stable. By contrast, consider $f(x) = x^2/2$. It is Morse at $p = 0$, with index $k = 0$. Take a perturbation $g(x) = x^2/2 - \epsilon x$ so that $dg = x - \epsilon$. At its singular point ϵ, $d^2 g = 1$ and the index remains $k = 0$.

In some sense, catastrophe theory is about the study of singularities (critical points) which are *not Morse*, and as with all subjects in this book, we are interested in the classification of how non-Morse (but still following some notion of stability) singularities can be.

We already saw that $f(x) = x^3$ is unstable. This belongs to an obvious family originally noted by Thom and given very creative names. The case of $f = x$ is non-singular and that of $f = x^2 + ax$ is a stable local quadratic minimum. Our example of $f = x^3 + ax$ is called the **fold**. And the higher powers $f = x^4 + ax^2 + bx$, $f = x^5 + ax^3 + bx^2 + cx$, and $f = x^6 + ax^4 + bx^3 + cx^2 + dx$ for real coefficients a, b, c, d are called the **cusp**, the **swallowtail** and the **butterfly**, respectively. Moreover, there is $f = x^n + \ldots$ for integer $n > 6$. A more precise statement of Thom's theorem requires two definitions.

DEFINITION 5.26 A germ of a function $f \colon \mathbb{R}^n \to \mathbb{R}$ at 0 is the equivalent class of functions which are identical on a neighbourhood of 0.

Here, the neighbourhood depends on f. One can roughly think of a germ as functions which share Taylor series up to some truncated order. One can extend the variables from $x_i \in \mathbb{R}^n$ to $t_j \in \mathbb{R}^k$ and define

DEFINITION 5.27 A k-parameter unfolding (also called a k-parameter family of germs based on f) of $f: \mathbb{R}^n \to \mathbb{R}$ is a germ $F: \mathbb{R}^k \times \mathbb{R}^n \to \mathbb{R}$ where $F(t_j = 0, x_i) = f(x_i)$.

There is clearly a notion of equivalence between unfoldings in that the variables t_j are defined only up to coordinate change. Thom's theorem gives the classification for universal, stable k-unfoldings:

THEOREM 5.28 *There are 7 stable, universal unfoldings with $k \le 4$:*

Germ	k	Catastrophe
x^3	*1*	*Fold*
$\pm x^4$	*2*	*Cusp*
x^5	*3*	*Swallow-tail*
$\pm x^6$	*4*	*Butterfly*
$x^3 + xy^3$ *or* $x^3 + y^3$	*3*	*Hyperbolic umbilic*
$x^3 - xy^2$	*3*	*Elliptic umbilic*
$\pm(x^2 y + y^4)$	*4*	*Parabolic umbilic*

Arnold extends this to all simple singularities, where simple means that all neighbouring singularities (perturbations) fall into a *finite* number of equivalent classes (again, finiteness is key):

THEOREM 5.29 *Simple stable singularities of $f: \mathbb{R}^n \to \mathbb{R}$ follow an ADE classification:*

$$
\begin{array}{c|c}
A_{k\ge 1} & x_1^{k+1} + x_2^2 + \sum_{i=3}^{n} x_i^2 \\[2ex]
D_{k\ge 4} & x_1(x_1^{k-2} + x_2^2) + \sum_{i=3}^{n} x_i^2 \\[2ex]
E_6 & x_1^4 + x_2^3 + \sum_{i=3}^{n} x_i^2 \\[2ex]
E_7 & x_2(x_1^3 + x_2^2) + \sum_{i=3}^{n} x_i^2 \\[2ex]
E_8 & x_1^5 + x_2^3 + \sum_{i=3}^{n} x_i^2
\end{array}
$$

In particular [150], we see that the fold, cusp, swallow-tail and butterfly are $A_{k=3,4,5,6}$ and the umbilics are $D_{k=4,5}$.

Thus, summarising, in the transmission of a front in a manifold without boundary the typical singularities are of ADE type. Though note that when going over to manifolds with boundary the list of typical singularities is widened to all crystallographic types, and when including obstacles then in the context of caustics the possible singularities also include those corresponding to the non-crystallographic groups H_2, H_3, H_4 [151].

5.8 Calabi–Yau: du Val Singularities

We now see that there is a complex version of the above Arnold story, which is, in fact, the geometric version of McKay's correspondence.

First, let us recall some key facts from invariant theory (cf. e.g., [152]), a field started by Noether and Hilbert, originally motivated by the study of polynomial invariants under finite group action (cf. degrees and exponents in Section 3.2.4). As always, we will minimise formal theory, and emphasise examples and classification patterns.

Take the discrete finite subgroups of SU(2), on which we dwelt in Section 3.4.2: these can be thought of as complex 2×2 matrices acting on $\mathbb{C}^2$ with coordinates (x, y). **Classical invariant theory** then prescribes a standard way to find the invariants under the group action explicitly as polynomials in (x, y) of a given degree (see also Section 3.2.4). A theorem of A. Noether ensures that the ring of invariants is finitely generated. To count the number a_i of independent invariants at degree i, the generating function is given by the Molien series

$$M(t; G) = \frac{1}{|G|} \sum_{g \in G} \frac{1}{\det(\mathbb{I} - tg)} = \sum_{i=0}^{\infty} a_i t^i \,, \tag{5.21}$$

which is always a rational function in t.

For example, take $\mathbb{Z}/2\mathbb{Z} = \langle \mathrm{Diag}(-1, -1) \rangle = \left\{ \left(\begin{smallmatrix} 1 & 0 \\ 0 & 1 \end{smallmatrix} \right), \left(\begin{smallmatrix} -1 & 0 \\ 0 & -1 \end{smallmatrix} \right) \right\} \subset$ SU(2). The non-identity element takes $(x, y) \mapsto (-x, -y)$. The independent monomial invariants, in increasing degree, are obviously

Degree	Invariants	# Invariants
0	$\{1\}$	1
1	$\{\}$	0
2	$\{x^2, xy, y^2\}$	3
3	$\{\}$	0
4	$\{x^4, xy^3, x^2y^2, x^3y, y^4\}$	5
$\cdots$	$\cdots$	$\cdots$

$$\tag{5.22}$$

The Molien series here is

$$M(t; \mathbb{Z}/2\mathbb{Z}) = \frac{1}{2}\left(\frac{1}{(1-t)^2} + \frac{1}{(1+t)^2}\right) = \frac{1+t^2}{(1-t^2)^2}, \qquad (5.23)$$

so that the Taylor coefficients are $a_{2i} = 2i + 1$ and $a_{2i-1} = 0$ for $i \in \mathbb{Z}_{>0}$, in agreement with the table above.

The above can be re-phrased geometrically, in the language of modern **algebraic geometry**.[11] The $\mathbb{C}^2$ on which G acts is an affine (complex) algebraic variety.[12] We now see that the quotient $\mathbb{C}^2/G$ is also an affine variety. This quotient is not smooth because the origin $(0, 0)$ is a fixed point. Such a singular space is called a V-manifold or **orbifold** [155]. Note that such spaces are not manifolds – the patch near the singularity is not homeomorphic to $\mathbb{C}^n$ – but are varieties. As such, they should be describable as vanishing loci of polynomials.

We illustrate how to obtain the polynomials by examining the ring of invariants: relations amongst the fundamental invariants will give the defining equations. In our above $\mathbb{Z}/2\mathbb{Z}$ example, there are the three *fundamental* invariants, coming from degree 2:

$$u := x^2, \ v := y^2, \ w := xy \quad \Rightarrow \quad uv = w^2. \qquad (5.24)$$

These are fundamental in that all the other invariants can be written as polynomials therein. Equation (5.24) tells us that the fundamental invariants obey a single algebraic relation: $uv = w^2$. Voilà, this is the defining equation for our $\mathbb{C}^2/(\mathbb{Z}/2\mathbb{Z})$ orbifold singularity, realised[13] as an affine algebraic variety in $\mathbb{C}^3$ with complex coordinates (u, v, w). It is a **hyper-surface** (a single defining polynomial) of complex dimension 2 (after all, it is a quotient of $\mathbb{C}^2$) living inside $\mathbb{C}^3$.

In this geometrical avatar, the Molien series of G is the Hilbert series.[14] In geometry, the Hilbert series is a key characterisation of an algebraic variety

---

[11] Introductions to algebraic geometry for the interested reader are, e.g., [153, 154].

[12] Basically, an algebraic variety is the vanishing locus of a system of polynomials. Complex affine means that all the variables are defined in $\mathbb{C}$, rather than some projective space. Here, $\mathbb{C}^2$ is parametrised by $(x, y) \in \mathbb{C}^2$ with no defining polynomial.

[13] In fancier algebro-geometric language, this realisation is the standard quotient-ring/algebraic-variety correspondence:

$$\mathbb{C}^2/(\mathbb{Z}/2\mathbb{Z}) \simeq \text{Spec}\left(\mathbb{C}[u, v, w]/\left\langle uv - w^2 \right\rangle\right),$$

with the affine variety established as a maximal spectrum – set of maximal ideals – of the polynomial ring in three variables.

[14] We recall that for a variety X in $\mathbb{C}[x_1, ..., x_k]$, the Hilbert series is the generating function for the dimension of the graded pieces: $H(t; X) = \sum_{i=-\infty}^{\infty}(\dim_{\mathbb{C}} X_i)t^i$, where X_i, the ith graded piece of X, can be thought of as the number of algebraically independent degree i polynomials

and its embedding into ambient space – a vast subject into which we sadly do not have space to go in detail.[15]

One can compute the defining polynomials – it turns out they are all hypersurfaces – for $\mathbb{C}^2/G$ for our finite ADE subgroups G of SU(2). For reference, we also include the explicit 2×2 matrix generators as well as the Molien/Hilbert series that were computed in Section 3.1.1 of [156] and Section 3.1 of [157]:

$G \subset \mathrm{SU}(2)$	$|G|$	Generators	*Equation*
$\widehat{A_{n-1}}$	n	$\langle R(n) \rangle$	$uv = w^n$
$\widehat{D_{n+2}}$	$4n$	$\langle R(2n), S_D \rangle$	$u^2 + v^2 w = w^{n+1}$
$\widehat{E_6}$	24	$\langle R_6, S, T_6 \rangle$	$u^2 + v^3 + w^4 = 0$
$\widehat{E_7}$	48	$\langle R_7, S, T_7 \rangle$	$u^2 + v^3 + vw^3 = 0$
$\widehat{E_8}$	120	$\langle R_8, S, T_8 \rangle$	$u^2 + v^3 + w^5 = 0$

$$(5.25)$$

on X. A useful property of $H(t)$ is that it is a rational function in t and can be written in two ways:

$$H(t; X) = \begin{cases} \frac{Q(t)}{(1-t)^k}, & \text{Hilbert series of the first kind;} \\ \frac{P(t)}{(1-t)^{\dim(X)}}, & \text{Hilbert series of the second kind.} \end{cases}$$

Importantly, both $P(t)$ and $Q(t)$ are polynomials with *integer* coefficients. The powers of the denominators are such that the leading pole captures the dimension of the embedding space $\mathbb{C}^k$ and of X, respectively.

[15] One fun thing to do with Hilbert/Molien series is to apply the plethystic programme [156, 157]. Now define, for some analytic function $f(t)$ which affords a Taylor series $\sum_{n=0}^{\infty} a_n t^n$, its plethystic exponential $PE[f(t)] := \exp\left(\sum_{n=1}^{\infty} \frac{f(t^n)-f(0)}{n}\right)$; then we have an Euler-type product formula $g(t) = PE[f(t)] = \prod_{n=1}^{\infty}(1 - t^n)^{-a_n}$. Furthermore, there is an explicit inverse, the plethystic logarithm, such that $f(t) = PE^{-1}(g(t)) = \sum_{k=1}^{\infty} \frac{\mu(k)}{k} \log(g(t^k))$, where $\mu(k)$ is the standard Möbius function for $k \in \mathbb{Z}_+$ which is 0 if k has repeated prime factors and is $(-1)^n$ if k factorises into n distinct primes, together with the convention that $\mu(1) = 1$.
A curious fact is that given Hilbert series $H(t; X)$ of an algebraic variety X, the plethystic logarithm is of the form $PE^{-1}[H(t; X)] = b_1 t + b_2 t^2 + b_3 t^3 + \ldots$ where all $b_n \in \mathbb{Z}$ and a positive b_n corresponds to a generator in the coordinate ring of X and a negative b_n to a relation. In particular, if X is a complete intersection, then $PE^{-1}[H(t; X)]$ is a finite polynomial. In our running example of $\mathbb{C}^2/(\mathbb{Z}/2\mathbb{Z})$, the plethystic logarithm gives $3t^2 - t^4$ exactly. This is in accord with the fact that the defining equation $uv - w^2 = 0$ in Equation (5.24) comes from 3 quadratic variables $(u, v, w) = (x^2, y^2, xy)$, obeying a single degree 4 relation as a hypersurface.

G	Molien/Hilbert series $M(t; G)$
$\widehat{A_{n-1}}$	$(1 + t^n)/(1 - t^2)(1 - t^n)$
$\widehat{D_{n+2}}$	$(1 + t^{2n+2})/(1 - t^4)(1 - t^{2n})$
$\widehat{E_6}$	$(1 - t^4 + t^8)/(1 - t^4 - t^6 + t^{10})$
$\widehat{E_7}$	$(1 - t^6 + t^{12})/(1 - t^6 - t^8 + t^{14})$
$\widehat{E_8}$	$\dfrac{1 + t^2 - t^6 - t^8 - t^{10} + t^{14} + t^{16}}{1 + t^2 - t^6 - t^8 - t^{10} - t^{12} + t^{16} + t^{18}}$

$$(5.26)$$

In the above equations, remembering the table in Equation (3.18), where we gave the abstract generators of the binary ADE groups, we here need the explicit 2×2 generators embedded in SU(2) in order to compute the Molien/Hilbert series, which we list here (note that the generator for $\widehat{A_{n-1}}$ is $R(n)$ rather than $R(2n)$ because the cyclic group is abelian and does not receive a non-trivial double cover). Defining $\omega_n := e^{\frac{2\pi i}{n}}$ to be the primitive nth root of unity, we also define

$$R(n) = \begin{pmatrix} \omega_n & 0 \\ 0 & \omega_n^{-1} \end{pmatrix}, \quad S_D = \underline{k} \, ;$$

$$R_6 = \underline{i}, \quad S = \frac{1}{2}\left(\mathbb{I}_{2\times 2} + \underline{i} + \underline{j} + \underline{k}\right), \quad T_6 = \frac{1}{2}\left(\mathbb{I}_{2\times 2} + \underline{i} + \underline{j} - \underline{k}\right),$$

$$R_7 = \frac{1}{\sqrt{2}}\left(\underline{i} + \underline{j}\right), \quad T_7 = \frac{1}{\sqrt{2}}\left(\mathbb{I}_{2\times 2} + \underline{i}\right),$$

$$R_8 = -(S T_8)^{-1}, \quad T_8 = \frac{1}{2}\left(\varphi\mathbb{I}_{2\times 2} + \varphi^{-1}\underline{i} + \underline{j}\right); \qquad (5.27)$$

where $\varphi = \frac{1}{2}(1 + \sqrt{5})$ is the golden ratio and $\underline{i}, \underline{j}$, and $\underline{k}$ are the quaternion basis (not the Pauli matrix basis)

$$\underline{i} = \begin{pmatrix} i & 0 \\ 0 & -i \end{pmatrix}, \quad \underline{j} = \begin{pmatrix} 0 & 1 \\ -1 & 0 \end{pmatrix}, \quad \underline{k} = \begin{pmatrix} 0 & i \\ i & 0 \end{pmatrix}. \qquad (5.28)$$

The list of hypersurfaces in (5.25) is very special; they are known as **du Val** singularities [158, 159], or Kleinian singularities. These should be compared to the Arnold singularities of Theorem 5.29.

Perhaps the easiest way to define du Val singularities is via Ricci-flatness. Essentially, because G is a subgroup of SU(2), orbifolds are special in that they are (locally) **Calabi–Yau**, using one of the many equivalent definitions of Calabi–Yau manifolds, in particular, that they are Kähler [50] and admit SU(n) holonomy [160]. We leave the reader to a modern introduction, especially in the context of theoretical physics, to Calabi–Yau manifolds in [24].

It suffices to mention here that Calabi–Yau[16] here means the metric is locally flat. In particular, for the $G = A_n$, cyclic, cases, the metric is called ALE, for asymptotically locally Euclidean [161]. In brief, a local or affine Calabi–Yau variety of complex dimension n with orbifold singularity is of the form

$$\mathbb{C}^n/\Gamma,$$

where Γ is a discrete, finite, subgroup of SU(n).

Now, dimension $n = 2$ is special, because *all* local Calabi–Yau varieties are of orbifold type. In other words, the classification theorem is that local Calabi–Yau 2-folds (also known as K3 surfaces),[17] are all of the form above [9]:

THEOREM 5.30 *All local singularities of Ricci-flat complex surfaces are du Val, i.e., orbifolds of the form $\mathbb{C}^2/G$ for G a discrete finite subgroup of SU(2), with defining equations given in Equation (5.25) as hypersurfaces in $\mathbb{C}^3$.*

How is this a geometric realisation of the McKay Correspondence? It turns out that one can resolve a singularity by a process called **blowup**, where one replaces the singular point at the origin with spheres (i.e., chains of $\mathbb{P}^1$s). The intersections amongst these spheres [9, 162] are encoded into intersection matrices. Lo and behold, these intersection matrices are precisely the Cartan matrices of the affine ADE algebras. In other words, local Calabi–Yau surface singularities are encoded by the affine $\widehat{ADE}$ Dynkin diagrams. The reader is also referred to [22, 163] for getting E_8 from the icosahedron using du Val singularities.

Indeed, generalising the McKay Correspondence to $n > 2$ is still an active area of research both in physics (as regards stringy Hodge numbers) [129, 164–166] and in mathematics (especially in algebraic geometry around derived equivalence) [167–176]; the next case of $n = 3$ is particularly interesting to string theorists because a cornerstone to compactification of 10-dimensional string theory is the set of Calabi–Yau threefolds (essentially because of the indisputable fact that $10 = 4 + 2 \times 3$).

---

[16] Purely algebro-geometrically, this means that the canonical sheaf on $\mathbb{C}^2/G$ admits a crepant resolution to a trivial canonical sheaf, and the singularity is called *canonical Gorenstein*. Because we are in complex dimension 2, the Calabi–Yau manifolds to which the singularities resolve are K3 surfaces.

[17] This is certainly not true in higher dimensions. Already for Calabi–Yau threefolds, there is a host of possible singular structures.

While we are on algebraic geometry, it is curious to point out the famous counter-example to Hilbert's 14th Problem [177] (see also [178]). Briefly, this was the question: given a field k and an intermediate field K between k and the field of rational functions in n variables with coefficients in k, i.e., $k \subset K \subset k(x_1, \ldots, x_n)$. Hilbert asked where the algebra of intersection $R := K \cap k[x_1, \ldots, x_n]$ is finitely generated over k. The original motivation came from invariant theory over $k = \mathbb{C}$ and $K = \mathbb{C}(x_1, \ldots, x_n)^G$, the field of invariant rational functions under a subgroup $G \subset GL(n, \mathbb{C})$. The counter-example of Nagata [179] and later generalised by Mukai [180] is as follows. Consider the action $x_i \to x_i$ and $y_i \to y_i + t_i x_i$ by group $\tilde{G}$ on $\mathbb{C}[x_1, \ldots, x_n; y_1, \ldots, y_n]$. Now, G is a vector space of dimension n. Mukai's generalisation – using T-shaped root systems – is that the ring of invariants $\mathbb{C}[x_1, \ldots, x_n; y_1, \ldots, y_n]^G$ for some co-dimension r subspace $G \subset \tilde{G}$ is *not* finitely generated if

$$\frac{1}{r} + \frac{1}{n-r} \leq \frac{1}{2}. \tag{5.29}$$

This is certainly reminiscent of our central equation (1.3) and there should be some underlying ADE story.

5.8.1 Some Modern Physics

When there is geometry, there is naturally physics. This geometric guise of the McKay Correspondence, via the Calabi–Yau nature, finds its role in string theory, in particular the AdS/CFT correspondence [128, 129, 181], in string compactifications on K3 surfaces [182, 183], as well as – albeit more mysteriously – conformal field theory [4, 24, 164, 181, 184–190].

Indeed, the canonical textbook [186] (based on [185], see a nice recent treatment in [188] which gives a reduction into the bare mathematical problem, as well as Theorem 6.2.2. in [4]) tells us the following (we will certainly not delve into the details of CFTs, for whose induction the interested reader is referred to the above references). Consider genus one partition functions of two-dimensional conformal field theories associated to the affine SU(2) algebra – i.e., $\widehat{A_1}$. Then, in terms of the characters χ (indexed by a combination of the highest weight and the central charge), the modular invariant partition functions are:

$$Z\left(A_{n_1;\, n\geq 3}\right) = \sum_{a=1}^{n-1} |\chi_a|^2 \,,$$

$$Z\left(D_{\frac{n}{2}+1;\, \frac{n}{2}\text{ even}}\right) = \sum_{a=1}^{n-1} \chi_a \chi_{n-2}^* \,,$$

$$Z\left(D_{\frac{n}{2}+1;\, \frac{n}{2}\text{ odd}}\right) = |\chi_1 + \chi_{n-1}|^2 + |\chi_3 + \chi_{n-3}|^2 + \cdots + 2|\chi_{\frac{n}{2}}|^2 \,,$$

$$Z(E_6) = |\chi_1 + \chi_7|^2 + |\chi_4 + \chi_8|^2 + |\chi_5 + \chi_{11}|^2 \,,$$

$$Z(E_7) = |\chi_1 + \chi_{17}|^2 + |\chi_5 + \chi_{13}|^2 + |\chi_7 + \chi_{11}|^2$$
$$+ \chi_9(\chi_3 + \chi_{15})^* + (\chi_3 + \chi_{15})\chi_9^* + |\chi_9|^2 \,,$$

$$Z(E_8) = |\chi_1 + \chi_{11} + \chi_{19} + \chi_{29}|^2 + |\chi_7 + \chi_{13} + \chi_{17} + \chi_{23}|^2 \,. \tag{5.30}$$

These clearly fall under an ADE pattern, which are so labelled presciently. This is the only ADE pattern that Terry Gannon did not consider reducible to a graph-theoretic/Perron–Frobenius-like problem in [4]. Understanding this phenomenon in more detail should lead to some interesting new mathematics and physics.

In the same vein, the fundamental equation (and related trichotomies in Equation (3.8) and Theorem 5.15, as well as the dichotomy in Theorem 3.9) of the McKay Correspondence in (4.12) has found its place in physics. Recall that, for a finite quiver with node labels n_i and adjacency matrix a_{ij}, we can consider, more generally,

$$kn_i - \sum_j a_{ij} n_j \,. \tag{5.31}$$

It turns out that this is proportional to the so-called beta-function in supersymmetric conformal field theories, especially those coming from string theory. When $k = 2$, the vanishing of Equation (5.31) gives the super-conformality condition for $\mathcal{N} = 2$ quiver gauge theory in physics, and the affine ADE Dynkin diagrams in mathematics. Implications of other values of k are discussed in [191] while [192] tri-partitions quiver theories as "good, bad, ugly" accordingly.

5.9 Elliptic Fibrations

Whilst we are on the topic of algebraic geometry, one cannot resist but go into Kodaira's ADE classification of elliptic fibrations. First, let us recall that a torus $T^2 = S^1 \times S^1$ (genus 1 Riemann surface) can be realised as a complex algebraic variety in $\mathbb{C}^2$, much like a circle S^1 is a real algebraic variety in $\mathbb{R}^2$ via the familiar $\{(x, y) \in \mathbb{R}^2 : x^2 + y^2 - 1 = 0\}$. In particular, the torus is a so-called elliptic curve, given by the vanishing locus of a cubic in $\mathbb{C}^2$:

$$\{(x, y) \in \mathbb{C}^2 : y^2 = 4x^3 - g_2 x - g_3\}, \tag{5.32}$$

where g_2 and g_3 are complex parameters that give the "shape" of the torus. Every cubic in 2 complex variables can be, via appropriate coordinate transformation, put into this canonical form, called the **Weierstraß form**. Isomorphism classes of an elliptic curve are given by the j-invariant, which we saw in Section 5.1.1. Explicitly in terms of the coefficients, $j := g_2^3/(g_2^3 - 27g_3^2)$.

Next, let us consider an elliptic fibration. Topologically, this is a variety which is locally a torus sitting on each point of a base. The simplest example, with which we will illustrate, is that of an elliptic surface (though the situation is more generally applicable), which is a complex surface S (i.e., a 4-dimensional real manifold) consisting of a *base*, which we can locally take to be the complex plane $\mathbb{C}$ (with coordinate w), on top of each point of which is a T^2, *the fibre*.

Algebraically,[18] this means that we can write the equation of S in Weierstraß form (5.32) as:

$$y^2 = 4x^3 - g_2(w)x - g_3(w) , \tag{5.33}$$

where g_2 and g_3 are arbitrary polynomials in the base coordinate w.

Thus, for an elliptic surface, the j-invariant becomes a meromorphic function in w, as a map from $\mathbb{C}^1$ to $\mathbb{C}^1$:

$$j(w) := \frac{g_2(w)^3}{\Delta(w)} , \qquad \Delta(w) := g_2(w)^3 - 27g_3(w)^2 . \tag{5.34}$$

When the denominator Δ, called **discriminant**, vanishes, the curve (elliptic fibre) becomes singular. This can be easily checked: consider the RHS of the Weierstraß form $f = 4x^3 - g_2 x - g_3$ and $f' = 12x^3 - g_2$. The elliptic curve becomes singular when $f = f' = 0$, which upon eliminating x, gives the condition $\Delta = 0$.

Kodaira [193] classified all possible singularity types by considering what happens to g_2 and g_3 on the singular locus $\Delta(w) = 0$. To give a trivial example, suppose $y^2 = 4x^3 - wx$ so that $g_2 = w$, $g_3 = 0$, and $\Delta = w^3$. The discriminant

---

[18] One usually works on *projective* varieties, so that the base is a $\mathbb{P}^1$ with projective coordinate w, rather than $\mathbb{C}^1$. The Weierstraß equation is then projectivised to be homogeneous cubic in $\mathbb{P}^2$ with projective coordinates $[x : y : z]$, as $zy^2 = 4x^3 - g_2 xz^2 - g_3 z^3$. Globally, we really should think of the complex coordinates (x, y, z) as being sections of the bundles $(L^{\oplus 2}, L^{\oplus 3}, L)$ where $L = O_{\mathbb{P}^1}(2)$ is the anti-canonical bundle of $\mathbb{P}^1$. What we work with in this section is the $z = 1$ affine patch for simplicity.

Δ vanishes at 0, so the singular locus is the point $\{w = 0\}$, where the order of vanishing of g_2 is 1, that of g_3 is infinity, and that of Δ, 3. A simple transformation renders the equation to be $(y/2)^2 = x(x^2 - w/4)$, which, upon redefining $y/2 = s$, $u = x$ and $v = x^2 - w/4$ gives the equation $s^2 = uv$, which is the surface singularity A_1 we saw in Section 5.8.

In general, massaging the Weierstraß form into surface singularity types is highly non-trivial and requires the so-called Tate–Nagell algorithm. The insight of Kodaira is the classification of the complete singularity types of the elliptic fibration. Defining $\mathrm{Ord}(f(w))$ as the order of vanishing of a function $f(w)$ at w_0, i.e., the Taylor series of $f(w)$ starts with $O(w - w_0)^{\mathrm{Ord}(f(w))}$, we have

THEOREM 5.31 (Kodaira)　*The elliptic fibration types are determined by the order of vanishing of g_2, g_3, Δ at the singular locus $\Delta = 0$, and are the following*

$\mathrm{Ord}(g_2)$	$\mathrm{Ord}(g_3)$	$\mathrm{Ord}(\Delta)$	*Kodaira notation*	*Singularity type*
≥ 0	≥ 0	0	*smooth*	$-$
0	0	n	I_n	A_{n-1}
≥ 1	1	2	II	*none*
≥ 1	≥ 2	3	III	A_1
≥ 2	2	4	IV	A_2
2	≥ 3	$n+6$	I_n^*	D_{n+4}
≥ 2	3	$n+6$	I_n^*	D_{n+4}
≥ 3	4	8	IV^*	E_6
3	≥ 5	9	III^*	E_7
≥ 4	5	10	II^*	E_8

Beautifully, we see another emergence of the ADE meta-pattern.[19]

5.10　Back to Arnold's Trinities

At the end of our journey into the ADE web, let us take stock. We have seen many ADE sets and connections between them. We have also seen individual cases of connections between exceptional cases such as H_3 and E_8 [21, 22], or the exceptionals (E_6, E_7, E_8) and sporadic (Monster, Baby Monster,

---

[19] In physics, this classification finds itself in the F-theory version of string theory. One can put F-theory on elliptically fibred Calabi–Yau manifolds to construct even-dimensional quantum field theories, and 7-branes correspond to different factors of the discriminant; cf. [194].

Fischer's Group). This latter connection is actually more in the spirit of a trinity than a full ADE correspondence. Arnold's trinities would make one surmise that the trinity $(\mathbb{R}, \mathbb{C}, \mathbb{H})$ may be involved. There have also been other attempts to generalise this connection between sporadic objects, e.g., the Freudenthal–Tits magic square attempts to include the octonions. But for the moment let us discuss the relationship between the Trinity and the countable families further.

In our first ADE example we have seen that sometimes the three exceptional cases (the symmetries of the Platonic solids) seem quite different from the infinite families (symmetries of the n-gons and prisms). In fact this is often the case. As we have already mentioned above, these three exceptional cases on their own have been termed by Arnold as a "Trinity" (i.e., they are the Trinity-part of the ADE classification, though he does not seem to have advocated this view), which he related to the three normed division algebras

$$(\mathbb{R}, \mathbb{C}, \mathbb{H}). \tag{5.35}$$

In fact, it may not always be obvious which countable families to include in an ADE set. We have seen one incarnation of this earlier in the context of Section 4.4 as well as the anecdote from one of the authors (PC) about two of his collaborators, Jaap Seidel and Jean-Marie Goethals, in the 1970s, in the context of working on graphs with smallest eigenvalue -2. Whilst it turned out that they had found the same three sporadic examples, they had found complementary families! That is, there were a total of two infinite families, and all in all the cases assembled into a full ADE set. We have seen in Section 4.4 that this motivation to find an ADE pattern can lead one to include an infinite family in formulating a conjecture that one might have excluded a priori.

But now that we have covered rather more advanced material over the course of the book, we can go back to the Trinity part in more detail: Arnold noticed patterns of three exceptional examples in many different areas of mathematics, and sought to relate these back to the fundamental Trinity of normed division algebras $(\mathbb{R}, \mathbb{C}, \mathbb{H})$. For some of the other trinities this is rather obvious, such as for the associated projective spaces

$$(\mathbb{R}P^n, \mathbb{C}P^n, \mathbb{H}P^n) \tag{5.36}$$

and in a related though less immediate way the spheres $(\mathbb{R}P^1 = S^1, \mathbb{C}P^1 = S^2, \mathbb{H}P^1 = S^4)$ with their associated Hopf bundles $(S^1 \to S^1, S^3 \to S^2, S^7 \to S^4)$. For others, e.g., our example of the Platonic Solids, the connection to $(\mathbb{R}, \mathbb{C}, \mathbb{H})$ is less (or not) obvious, e.g., why the octahedron should be a complexified version of the tetrahedron or the icosahedron a quaternionic analogue of the tetrahedron.

Arnold listed many examples of suggested Trinities and various connections between them in [2, 61], such as (tetrahedron, octahedron and icosahedron), (A_3, B_3, H_3), (D_4, F_4, H_4), (E_6, E_7, E_8) and many more. The topics and connections that he mentions are closely related to topics that we have also discussed, as his Trinity parts are the E-type exceptional cases in many of our ADE sets. Some connections are straightforward, such as with the cluster of trinities around Platonic solids, root systems, Coxeter groups, binary icosahedral groups $(2T, 2O, 2I)$ and their orders $(24, 48, 120)$, as well as number of roots in the root system $(12, 18, 30)$, sum of the dimensions of the irreps of the binary groups and triples of rotational symmetry orders $(233, 234, 235)$ which are all plausibly connected to each other. The connection between (A_3, B_3, H_3) and (D_4, F_4, H_4) was only noted by Arnold very indirectly, and we have seen above that it is indeed a non-trivial argument, and in fact encompasses a whole ADE pattern.

We also saw above that there is another mysterious trinity which one can add to the story. It is well-known that in the classification of finite simple groups [71, 72], there are (1) the alternating groups $\mathfrak{A}_{n\geq 5}$; (2) the Lie groups defined over finite fields; and (3) 26 exceptional cases called the Sporadics. Whilst this is not in any obvious ADE pattern, it was an old speculation of McKay whether the largest three, viz.,

$$\text{(Monster, Baby Monster, Fischer's Group)} \qquad (5.37)$$

might furnish the trinity that relates to (E_6, E_7, E_8). It is not clear if this could be extended to a full ADE pattern too.

A good way of thinking about Trinities is perhaps that it is easier to find three exceptional examples in different areas of mathematics, which can serve as inspiration and a bootstrap to try to uncover fuller ADE sets/correspondences by including appropriate infinite families. It is not clear which Trinities may form part of an ADE set and which don't. We expect that investigations into this mysterious web of correspondences will lead to a lot of very interesting mathematics. What will bridge what now seem different areas of mathematics will probably crystallise as beautiful and unifying objects in their own right. We would like to encourage the readers of this book – and the next generation of mathematicians and physicists – to use their creativity and start to forge paths into this new Polymathematics terrain.

5.11 Summarising Outlook: A Rogues' Gallery

We conclude this book with a list of ADE and affine ADE sets that we have encountered, from elementary to highly advanced topics, and with various connections amongst them:

(i) Platonic solids, prisms, polygons
(ii) Polyhedral groups
(iii) Coxeter groups / full polyhedral groups
(iv) Root systems
(v) Simply-laced diagrams / graphs (Smith)
(vi) Affine diagrams / graphs (Smith)
(vii) Tesselations of the plane
(viii) Binary polyhedral groups
(ix) ADE root systems
(x) ADE Lie algebras
(xi) Affine Lie algebras
(xii) Real Singularities (Arnold)
(xiii) Complex Singularities (du Val)
(xiv) Conformal field theory and modular invariants (WZW Models)
(xv) ALE spaces (Kronheimer)
(xvi) Elliptic fibrations (Kodaira)
(xvii) Quivers (Gabriel)
(xviii) Cluster algebras (Fomin and Zelevinsky)
(xix) Stringy Hodge numbers (Batyrev–Dais)
(xx) McKay Correspondence as Derived Equivalence (Bridgeland–King–Reid)

We hope that readers of this book will populate this web of connections with some of the missing links between different areas of mathematics.

Exercises

5.1 Calculate the possible Niemeier root systems so that their ADE root system building blocks satisfy the rank 24 condition as well as all building blocks having the same Coxeter number.

5.2 Find a representation of the graph of Figure 5.1 as a subset of the root system D_n for suitable n. Extend your construction to any generalised line graph. (Hint: reverse engineer Theorem 5.9.)

5.3 Show that a graph with least eigenvalue -2 which is represented by a subset of the root system A_n is the line graph of a bipartite graph.

5.4
- Show that a regular generalised line graph is either a line graph or a cocktail party graph.
- Show that a line graph $L(\mathcal{G})$ is regular if and only if either $\mathcal{G}$ is regular, or $\mathcal{G}$ is a *semiregular bipartite graph* (that is, vertices in the same bipartite block have the same valency).

5.5 Describe all subsets S of root systems such that $\langle \alpha | \beta \rangle \geq 0$ for all $\alpha, \beta \in S$.

- If the root system is decomposable, then S is the disjoint union of subsets of the indecomposable components, so it suffices to solve the problem for these.
- If the root system is of ADE type, then Theorem 5.9 describes such sets: if the type is A_n, then S is the line graph of a bipartite graph; if type D_n, then S is a generalised line graph; if type E_n, then the number of possibilities is finite.
- Consider types B_n and C_n. The root system is the disjoint union of a type D_n root system and an orthonormal basis, one of these scaled by a factor $\sqrt{2}$. In the D_n root system, we have a generalised line graph (which may not span the whole space); we may add vectors from the orthonormal basis if either they are not in the span of the generalised line graph or they are of vertex type in the proof of Theorem 5.9.
- In type $I_2(n)$, the set must be contained within a right-angled sector in the plane.
- The remaining cases, F_4, H_3 and H_4, contain only finitely many possibilities.

5.6 (Computing project) Determine the graphs represented by a subset of E_8 which are not generalised line graphs.

5.7 Show that a mutation is an involution (e.g., doing it twice returns us to the starting quiver).

5.8 Show that any orientation of the edges of a tree can be obtained from any other by a sequence of mutations.

References

[1] Y.-H. He, John Keith Stuart McKay: 1939–2022, obituary for the London Mathematical Society (2023). arXiv:2305.00850.

[2] V. I. Arnol'd, Mathematics: Frontiers and perspectives; Polymathematics: is mathematics a single science or a set of arts, American Mathematical Society, 2000.

[3] J. McKay, Graphs, singularities, and finite groups, Proceedings of the Symposium in Pure Mathematics 37 (1980) 183–186.

[4] T. Gannon, Moonshine beyond the Monster: The bridge connecting algebra, modular forms and physics, Cambridge University Press, 2006.

[5] S. P. Sirag, ADEX theory: How the ADE Coxeter graphs unify mathematics and physics, Vol. 57, World Scientific, 2016.

[6] T. Gannon, Monstrous moonshine and the classification of CFT, arXiv preprint math/9906167.

[7] F. E. Browder, Mathematical developments arising from Hilbert problems, American Mathematical Society, 1976.

[8] N. Hitchin, Clay academy lecture, April 2005.
www.claymath.org/library/academy/LectureNotes05/Hitchin.pdf.

[9] P. Slodowy, Simple singularities and simple algebraic groups, Vol. 815, Springer, 2006.

[10] D. N. Marshall, Carved stone balls, Proceedings of the Society of Antiquaries of Scotland Edinburgh 108 (1976) 40–72.

[11] J. Stillwell, The story of the 120-cell, Notices of the AMS 48 (1) (2001) 17–24.

[12] J.-P. Luminet, J. R. Weeks, A. Riazuelo, R. Lehoucq, J.-P. Uzan, Dodecahedral space topology as an explanation for weak wide-angle temperature correlations in the cosmic microwave background, Nature 425 (6958) (2003) 593–595.

[13] I. Polo-Blanco, Alicia Boole Stott, a geometer in higher dimension, Historia Mathematica 35 (2) (2008) 123–139.

[14] M. Artin, Algebra, Prentice-Hall, 1991.

[15] J. H. Conway, H. Burgiel, C. Goodman-Strauss, The symmetries of things, CRC Press, 2016.

[16] P.-P. Dechant, Clifford algebra is the natural framework for root systems and Coxeter groups. Group theory: Coxeter, conformal and modular groups, Advances in Applied Clifford Algebras 27 (1) (2015) 17–31.

[17] H. S. M. Coxeter, Regular polytopes, Dover books on advanced mathematics, Dover Publications, 1973.

[18] H. S. M. Coxeter, Discrete groups generated by reflections, Annals of Mathematics 35 (1934) 588–621.

[19] H. S. M. Coxeter, The complete enumeration of finite groups of the form $r_i^2 = (r_i r_j)^{k_{ij}} = 1$, Journal of the London Mathematical Society s1-10 (1) (1935) 21–25.

[20] J. E. Humphreys, Reflection groups and Coxeter groups, Vol. 29 of Cambridge studies in advanced mathematics, Cambridge University Press, 1992.

[21] P.-P. Dechant, The birth of E_8 out of the spinors of the icosahedron, Proceedings of the Royal Society A 20150504.

[22] J. C. Baez, From the icosahedron to E_8, arXiv preprint arXiv:1712.06436.

[23] J. E. Humphreys, Reflection groups and Coxeter groups, Cambridge University Press, 1990.

[24] Y.-H. He, The Calabi-Yau landscape: from geometry, to physics, to machine-learning, Springer, LNM 2293, 2018.

[25] J. Patera, R. Twarock, Affine extensions of noncrystallographic Coxeter groups and quasicrystals, Journal of Physics A: Mathematical and General 35 (2002) 1551–1574.

[26] P.-P. Dechant, C. Boehm, R. Twarock, Novel Kac-Moody-type affine extensions of non-crystallographic Coxeter groups, Journal of Physics A: Mathematical and Theoretical 45 (28) (2012) 285202.

[27] P.-P. Dechant, C. Boehm, R. Twarock, Affine extensions of non-crystallographic Coxeter groups induced by projection, Journal of Mathematical Physics 54 (9) (2013) 093508.

[28] V. G. Kac, Infinite-dimensional Lie algebras, Vol. 44 of Progress in mathematics, Cambridge University Press, 1994.

[29] K. Becker, M. Becker, J. H. Schwarz, String theory and M-theory, Cambridge University Press, 2007.

[30] N. Berkovits, J. Maldacena, $N = 2$ superconformal description of superstring in Ramond-Ramond plane wave backgrounds, Journal of High Energy Physics 2002 (10) (2002) 059.

[31] P. J. Cameron, Combinatorics: topics, techniques, algorithms, Cambridge University Press, 1994.

[32] Y.-H. He, S.-T. Yau, Graph Laplacians, Riemannian manifolds and their machine-learning, Mathematics, Computation and Geometry of Data 2 (2022) 1–48.

[33] J. H. Smith, Some properties of the spectrum of a graph, Proceeding of International Conference Combinatorial Structures and Their Applications (1970) 403–406.

[34] N. Bourbaki, Groupes et algèbres de Lie, chapitres 4, 5 et 6, Masson, 1981.

[35] P. J. Cameron, J.-M. Goethals, J. J. Seidel, E. E. Shult, Line graphs, root systems, and elliptic geometry, Journal of Algebra 43 (1976) 305–327.

[36] D. Hestenes, Space-time algebra, Gordon and Breach, 1966.

[37] C. Doran, A. N. Lasenby, Geometric algebra for physicists, Cambridge University Press, 2003.

[38] I. R. Porteous, Clifford algebras and the classical groups, Cambridge University Press, 1995.

[39] P. Lounesto, Clifford algebras and spinors, 2nd ed., London mathematical society lecture note series, Cambridge University Press, 2001.

[40] D. J. H. Garling, Clifford algebras: an introduction, London mathematical society student texts, Cambridge University Press, 2011.

[41] D. Hestenes, New foundations for classical mechanics; 2nd ed., Fundamental theories of physics, Kluwer, 1999.

[42] D. Hestenes, J. W. Holt, The Crystallographic Space Groups in Geometric Algebra, Journal of Mathematical Physics 48 (2007) 023514.

[43] P. A. M. Dirac, Wave equations in conformal space, The Annals of Mathematics 37 (2) (1936) 429–442.

[44] D. Hestenes, G. Sobczyk, Clifford algebra to geometric calculus: a unified language for mathematics and physics, in: Fundamental theories of physics, Reidel, 1984.

[45] P.-P. Dechant, A 3d spinorial view of 4d exceptional phenomena, in: Symmetries in graphs, maps, and polytopes workshop, Springer International Publishing, 2014, pp. 81–95.

[46] P.-P. Dechant, Clifford algebra unveils a surprising geometric significance of quaternionic root systems of Coxeter groups, Advances in Applied Clifford Algebras 23 (2) (2013) 301–321.

[47] P.-P. Dechant, Clifford spinors and root system induction: H_4 and the grand antiprism, Advances in Applied Clifford Algebras 31 (2021).

[48] T. Frankel, The geometry of physics: an introduction, Cambridge University Press, 2011.

[49] J. Stewart, J. M. Stewart, Advanced general relativity, Cambridge University Press, 1993.

[50] A. Moroianu, Lectures on Kähler geometry, Vol. 69, Cambridge University Press, 2007.

[51] W. Fulton, J. Harris, Representation theory: A first course, Springer-Verlag, 1991.

[52] J. E. Humphreys, Introduction to Lie algebras and representation theory, Vol. 9, Springer Science & Business Media, 2012.

[53] J. Fuchs, C. Schweigert, Symmetries, Lie algebras and representations, Cambridge University Press, 2003.

[54] J. E. Humphreys, Introduction to Lie algebras and representation theory, Vol. 9, Springer Science & Business Media, 2012.

[55] Y.-H. He, J. McKay, Sporadic and exceptional, arXiv preprint arXiv:1505.06742.

[56] V. G. Kac, Infinite-dimensional Lie algebras, Cambridge University Press, 1990.

[57] T. Keef, R. Twarock, Affine extensions of the icosahedral group with applications to the three-dimensional organisation of simple viruses, Journal of mathematical biology 59 (2009) 287–313.

[58] P.-P. Dechant, C. Boehm, R. Twarock, Novel Kac-Moody-type affine extensions of non-crystallographic Coxeter groups, Journal of Physics A: Mathematical and Theoretical 45 (28) (2012) 285202. arXiv:1110.5219.

[59] P.-P. Dechant, C. Boehm, R. Twarock, Affine extensions of non-crystallographic coxeter groups induced by projection, Journal of Mathematical Physics 54 (9) (2013).

[60] P.-P. Dechant, Rank-3 root systems induce root systems of rank 4 via a new Clifford spinor construction, Journal of Physics: Conference Series 597 (1) (2015) 12027.

[61] V. I. Arnold, Symplectization, complexification and mathematical trinities, Fields Institute Communications 24 (1999) 23–37.

[62] P.-P. Dechant, Platonic solids generate their four-dimensional analogues, Acta Crystallographica Section A: Foundations of Crystallography 69 (6) (2013) 592–602.

[63] P.-P. Dechant, From the Trinity (A_3, B_3, H_3) to an ADE correspondence, Proceedings of the Royal Society A 474 (2220) (2018) 20180034.

[64] P.-P. Dechant, The E_8 geometry from a Clifford perspective, Advances in Applied Clifford Algebras 27 (1) (2017) 397–421.

[65] P.-P. Dechant, A Clifford algebraic framework for Coxeter group theoretic computations, Advances in Applied Clifford Algebras 24 (1) (2014) 89–108.

[66] S. Chen, P.-P. Dechant, Y.-H. He, E. Heyes, E. Hirst, D. Riabchenko, Machine learning Clifford invariants of ADE Coxeter elements, Advances in Applied Clifford Algebras 34 (20) (2024).

[67] W. Shakespeare, The Winter's Tale: Third Series, Vol. 44, A&C Black, 2010.

[68] J. H. Conway, N. J. A. Sloane, E. Bannai, Sphere-packings, lattices, and groups, Springer-Verlag, 1987.

[69] R. A. Wilson, The geometry of the Hall-Janko group as a quaternionic reflection group, Geometriae Dedicata 20 (1986) 157–173.

[70] R. V. Moody, J. Patera, The E_8 family of quasicrystals, in: Proceedings NATO ASI Noncompact Lie Groups and Some of Their Applications, San Antonio, eds. Tanner, E.A. and Wilson, F., NATO ASI Series C 429 (1993) 341.

[71] D. Gorenstein, Finite groups, Vol. 301, American Mathematical Soc., 2007.

[72] J. H. Conway, R. Curtis, S. Norton, R. Parker, R. Wilson, Atlas of finite groups. maximal subgroups and ordinary characters for simple groups. with computational assistance from J.G. Thackray (1985).

[73] R. W. Carter, Simple groups of Lie type, reprint of the 1972 orig. Edition John Wiley & Sons, Inc., 1989.

[74] J. H. Conway, S. P. Norton, Monstrous moonshine, Bulletin of the London Mathematical Society 11 (3) (1979) 308–339.

[75] R. E. Borcherds, Monstrous moonshine and monstrous Lie superalgebras, Inventiones mathematicae 109 (1) (1992) 405–444.

[76] V. G. Kac, Infinite-dimensional algebras, Dedekind's η-function, classical Möbius function and the very strange formula, Advances in Mathematics 30 (2) (1978) 85–136.

[77] Y.-H. He, J. McKay, Eta Products, BPS States and K3 Surfaces, JHEP 01 (2014) 113.

[78] G. Glauberman, S. P. Norton, On McKay's connection between the affine E_8 diagram and the Monster, in: Proceedings on Moonshine and related topics (Montreal, QC, 1999), 2001, pp. 37–42.

[79] C. H. Lam, H. Yamada, H. Yamauchi, Vertex operator algebras, extended E_8 diagram, and McKay's observation on the Monster simple group, Transactions of the American Mathematical Society 359 (9) (2007) 4107–4123.

[80] J. F. Duncan, Arithmetic groups and the affine E_8 Dynkin diagram, Groups and Symmetries: From Neolithic Scots to John McKay 47 (2009) 135.

[81] G. Höhn, C. H. Lam, H. Yamauchi, McKay's E_7 observation on the Baby Monster, International Mathematics Research Notices 2012 (1) (2012) 166–212.

[82] G. Höhn, C. H. Lam, H. Yamauchi, McKay's E_7 and E_6 observations on the Babymonster and the largest Fischer group, arXiv preprint arXiv:1002.1777.

[83] T. Eguchi, H. Ooguri, Y. Tachikawa, Notes on the K3 surface and the Mathieu group M_{24}, Experimental Mathematics 20 (1) (2011) 91–96.

[84] J. H. Bruinier, G. Van der Geer, G. Harder, D. Zagier, The 1-2-3 of modular forms, Springer Science & Business Media, 2008.

[85] D. Zagier, Ramanujan's mock theta functions and their applications (d'apres Zwegers and Ono-Bringmann), Astérisque 326 (2009) Séminaire Bourbaki 143–164.

[86] M. R. Gaberdiel, S. Hohenegger, R. Volpato, Mathieu moonshine in the elliptic genus of K3, Journal of High Energy Physics 2010 (10) (2010) 62.

[87] T. Creutzig, G. Hoehn, Mathieu moonshine and the geometry of K3 surfaces, arXiv preprint arXiv:1309.2671.

[88] M. Cheng, X. Dong, J. Duncan, J. Harvey, S. Kachru, T. Wrase, Mathieu Moonshine and $\mathcal{N} = 2$ string compactifications, arXiv preprint arXiv:1306.4981.

[89] M. C. Cheng, K3 surfaces, $\mathcal{N} = 4$ dyons, and the Mathieu group M_{24}, arXiv preprint arXiv:1005.5415.

[90] A. Chattopadhyaya, J. R. David, Gravitational couplings in $\mathcal{N} = 2$ string compactifications and Mathieu moonshine, arXiv preprint arXiv:1712.08791.

[91] A. Banlaki, A. Chowdhury, A. Kidambi, M. Schimpf, H. Skarke, T. Wrase, Calabi-Yau manifolds and sporadic groups, Journal of High Energy Physics 2018 (2) (2018) 1–35.

[92] K. Bringmann, J. Duncan, L. Rolen, Maass-Jacobi Poincaré series and Mathieu moonshine, arXiv preprint arXiv:1409.4124.

[93] A. Taormina, K. Wendland, A twist in the M_{24} moonshine story, arXiv preprint arXiv:1303.3221.

[94] A. Taormina, K. Wendland, The overarching finite symmetry group of Kummer surfaces in the Mathieu group M_{24}, arXiv preprint arXiv:1107.3834.

[95] A. Taormina, K. Wendland, The symmetries of the tetrahedral Kummer surface in the Mathieu group M_{24}, arXiv preprint arXiv:1008.0954.

[96] H.-V. Niemeier, Definite quadratische Formen der Dimension 24 und Diskriminante 1, Journal of Number Theory 5 (2) (1973) 142–178.

[97] M. C. Cheng, J. F. Duncan, J. A. Harvey, Umbral moonshine, arXiv preprint arXiv:1204.2779.

[98] J. F. Duncan, J. A. Harvey, The umbral moonshine module for the unique unimodular niemeier root system, arXiv preprint arXiv:1412.8191.

[99] J. A. Harvey, S. Murthy, C. Nazaroglu, ADE double scaled little string theories, mock modular forms and umbral moonshine, arXiv preprint arXiv:1410.6174.

[100] V. Anagiannis, M. C. Cheng, S. M. Harrison, K3 elliptic genus and an umbral moonshine module, arXiv preprint arXiv:1709.01952.

[101] M. Cheng, D. Whalen, Generalised umbral moonshine, arXiv preprint arXiv:1608.07835.

[102] V. Anagiannis, M. C. Cheng, TASI lectures on moonshine, arXiv preprint arXiv:1807.00723.

[103] Y. Hou, Elliptic genera of ADE type singularities, Ph.D. thesis, Dissertation, Universität Freiburg, 2021 (2020).

[104] J. F. Duncan, M. J. Griffin, K. Ono, Proof of the umbral moonshine conjecture, arXiv preprint arXiv:1503.01472.

[105] M. C. Cheng, S. Harrison, Umbral moonshine and K3 surfaces, arXiv preprint, arXiv:1406.0619.

[106] J. Tits, Buildings of spherical type and finite BN-pairs, Vol. 386 of Lect. Notes Math., Springer, 1974.

[107] E. E. Shult, Characterizations of certain classes of graphs, Journal of Combinatorial Theory, Series B 13 (1972) 142–167.

[108] F. Buekenhout, E. Shult, On the foundations of polar geometry, Geometriae Dedicata 3 (1974) 155–170.

[109] P. Erdős, A. Rényi, V. T. Sós, On a problem of graph theory, Studia Scientiarum Mathematicarum Hungarica 1 (1966) 215–235.

[110] P. J. Cameron, J. H. van Lint, Designs, graphs, codes and their links, Vol. 22 of London mathematical society student texts, Cambridge University Press, 1991.

[111] N. Biggs, Algebraic graph theory, no. 67 in Cambridge mathematical library, Cambridge University Press, 1993.

[112] A. J. Hoffman, C. A. Micchelli, Selected papers of Alan Hoffman with commentary, World Scientific, 2003.

[113] F. Bussemaker, D. Cvetkovic, J. J. Seidel, Graphs related to exceptional root systems, in: Geometry and combinatorics, Elsevier, 1991, pp. 94–100.

[114] D. Cvetkovic, P. Rowlinson, S. Simic, Spectral generalizations of line graphs: on graphs with least eigenvalue-2, Vol. 314, Cambridge University Press, 2004.

[115] A. J. Hoffman, On graphs whose least eigenvalue exceeds $-1 - \sqrt{2}$, Linear Algebra and Its Applcations 16 (1977) 153–165.

[116] H. Whitney, Congruent graphs and the connectivity of graphs, Hassler Whitney Collected Papers, American Journal 54 (1932) 150–168.

[117] P. J. Cameron, A note on generalized line graphs, Journal of Graph Theory 4 (2) (1980) 243–245.

[118] I. M. Gel'fand, V. Ponomarev, Quadruples of subspaces of a finite-dimensional vector space, in: Doklady Akademii Nauk, Vol. 197, 4, Russian Academy of Sciences, 1971, pp. 762–765.

[119] L. A. Nazarova, S. A. Ovsienko, A. V. Roiter, Polyquivers of finite type, Trudy Matematicheskogo Instituta imeni VA Steklova 148 (1978) 190–194.

[120] P. Donovan, M. R. Freislich, The representation theory of finite graphs and associated algebras, Carleton Mathematical Lecture Notes, issue 5, Carleton University, 1973.

[121] P. Gabriel, Unzerlegbare Darstellungen I, Manuscripta Mathematica 6 (1972) 71–103.

[122] P. Gabriel, Indecomposable representations II, Symposia Mathematica XI (1973) 81–104.

[123] Y. Drozd, Tame and wild matrix problems, *Thematic Work Collective, and Kiev* 1977 (1977) 104–114, (English translation: *Translation, Ser. 2, American Mathematic Society* 128 (1986) 31–55.

[124] L. Nazarova, S. Ovsienko, A. Roiter, Polyquivers of a finite type (Russian), Akad. Nauk Ukrain. SSR Inst. 23 (1977) 17–23.

[125] L. Nazarova, Representations of polyquivers of tame type (Russian), Otdel. Mat. Inst. Steklov. (LOMI) 286 (1977) 181–206.

[126] L. Nazarova, Polyquivers of infinite type (Russian), Trudy Mat. Inst. Steklov. 275 (19778) 175–189.

[127] D. Simson, Linear Representations of Partially Ordered Sets and Vector Space Categories, Vol. Algebra, Logic and Applications Series, 4, Gordon and Breach Science Publishers, 1992.

[128] M. R. Douglas, G. W. Moore, D-branes, quivers, and ALE instantons, arXivarXiv:hep-th/9603167.

[129] A. Hanany, Y.-H. He, Nonabelian finite gauge theories, JHEP 2 (1999) 13.

[130] Y.-H. He, Quiver Gauge theories: Finitude and trichotomy, Mathematics 6 (12) (2018) 291.

[131] S. Fomin, A. Zelevinsky, Cluster algebras I: foundations, Journal of the American Mathematical Society 15 (2) (2002) 497–529.

[132] S. Fomin, A. Zelevinsky, Cluster algebras II: Finite type classification, Inventiones Mathematicae 154 (1) (2003) 63–121.

[133] B. R. Marsh, Lecture notes on cluster algebras, AMC 10 (2014) 12.

[134] S. Fomin, L. Williams, A. Zelevinsky, Cluster Algebras, online, to appear. http://people.math.harvard.edu/~williams/book.html

[135] S. Fomin, N. Reading, Root systems and generalized associahedra, arXiv preprint, arXiv:math/0505518.

[136] G. Musiker, C. Stump, A compendium on the cluster algebra and quiver package in sage (2011). arXiv:1102.4844.

[137] P.-P. Dechant, Y.-H. He, E. Heyes, E. Hirst, Cluster algebras: Network science and machine learning, Journal of Computational Algebra 8 (2023) 100008.

[138] M.-W. Cheung, P.-P. Dechant, Y.-H. He, E. Heyes, E. Hirst, J.-R. Li, Clustering cluster algebras with clusters, Advances in Theoretical and Mathematical Physics, arXiv preprint arXiv:2212.09771 27 (3).

[139] B. Feng, A. Hanany, Y.-H. He, A. M. Uranga, Toric duality as Seiberg duality and brane diamonds, JHEP 12 (2001) 35.

[140] B. Feng, A. Hanany, Y.-H. He, D-brane gauge theories from toric singularities and toric duality, Nuclear Physics B 595 (2001) 165–200.

[141] V. Chandrasekaran, R. Longo, G. Penington, E. Witten, An algebra of observables for de sitter space, Journal of High Energy Physics 2023 (No. 2, Paper No. 82, p. 56).

[142] V. F. Jones, Index for subfactors, Inventiones mathematicae 72 (1) (1983) 1–25.

[143] V. Jones, S. Morrison, N. Snyder, The classification of subfactors of index at most 5, Bulletin of the American Mathematical Society 51 (2) (2014) 277–327.

[144] Y. Lin, L. Lu, S.-T. Yau, Ricci curvature of graphs, Tohoku Mathematical Journal, Second Series 63 (4) (2011) 605–627.

[145] Y. Lin, S.-T. Yau, A brief review on geometry and spectrum of graphs, arXiv preprint arXiv:1204.3168.

[146] R. Thom, Structural stability and morphogenesis, CRC Press, 2018.

[147] E. C. Zeeman, Catastrophe theory, in: W. Gttinger, H. Eikemeier (eds.) Structural stability in physics, Proceedings of two international symposia on applications of catastrophe theory and topological concepts in physics, Tbingen, Fed. Rep. of Germany, May 2–6 and December 11–14, 1978. (English) Zbl 0404.00015. Springer Series in Synergetics, Vol. 4. Springer-Verlag, 1979, pp. 12–22.

[148] M. Golubitsky, An introduction to catastrophe theory and its applications, SIAM Review 20 (2) (1978) 352–387.

[149] V. I. Arnol'd, Catastrophe theory, Springer Science & Business Media, 2003.

[150] V. I. Arnol'd, Normal forms for functions near degenerate critical points, the Weyl groups of A_k, D_k, E_k and Lagrangian singularities, Funktsional'nyi Analiz i ego Prilozheniya 6 (4) (1972) 3–25.

[151] O. P. Shcherbak, Wavefronts and reflection groups, Russian Mathematical Surveys 43 (3) (1988) 149.

[152] D. Eisenbud, The geometry of syzygies: a second course in algebraic geometry and commutative algebra, Vol. 229, Springer Science & Business Media, 2005.

[153] M. Reid, Undergraduate algebraic geometry, no. 12 in London mathematical society student texts, Cambridge University Press, 1988.

[154] B. Hassett, Introduction to algebraic geometry, Cambridge University Press, 2007.

[155] I. Satake, On a generalization of the notion of manifold, Proceedings of the National Academy of Sciences 42 (6) (1956) 359–363.

[156] S. Benvenuti, B. Feng, A. Hanany, Y.-H. He, Counting BPS operators in gauge theories: quivers, syzygies and plethystics, Journal of High Energy Physics 2007 (11) (2007) 50.

[157] B. Feng, A. Hanany, Y.-H. He, Counting gauge invariants: the plethystic program, Journal of High Energy Physics 2007 (03) (2007) 090.

[158] P. Du Val, On isolated singularities of surfaces which do not affect the conditions of adjunction (part I), in: Mathematical proceedings of the Cambridge philosophical society, Vol. 30, 4, Cambridge University Press, 1934, pp. 453–459.

[159] P. D. Val, Homographies, quaternions, and rotations, Oxford mathematical monographs, Clarendon Press, 1964.

[160] S.-T. Yau, On the Ricci curvature of a compact Kähler manifold and the complex Monge-Ampère equation, I, Communications on Pure and Applied Mathematics 31 (3) (1978) 339–411.

[161] P. B. Kronheimer, H. Nakajima, Yang-Mills instantons on ALE gravitational instantons, Mathematische Annalen 288 (1) (1990) 263–307.

[162] D. Huybrechts, Lectures on K3 surfaces, Vol. 158, Cambridge University Press, 2016.

[163] J. C. Baez, The octonions, Bulletin of the American Mathematical Society 39 (2002) 145–205.

[164] Y.-H. He, J. S. Song, Of McKay correspondence, nonlinear sigma model and conformal field theory, Advances in Theoretical and Mathematical Physics 4 (2000) 747–790.

[165] P. Mayr, Phases of supersymmetric D-branes on Kähler manifolds and the McKay correspondence, Journal of High Energy Physics 2001 (01) (2001) 018.

[166] G. Moore, A. Parnachev, Localized tachyons and the quantum McKay correspondence, Journal of High Energy Physics 2004 (11) (2005) 086.

[167] Y. Ito, M. Reid, The McKay correspondence for finite subgroups of $SL(3, C)$, Higher-Dimensional Complex Varieties (Trento, 1994) (1994) 221–240.

[168] V. V. Batyrev, D. I. Dais, Strong McKay correspondence, string-theoretic Hodge numbers and mirror symmetry, Topology 35 (4) (1996) 901–929.

[169] M. Reid, McKay correspondence, arXiv preprint, arXiv:alg-geom/9702016.

[170] T. Bridgeland, A. King, M. Reid, Mukai implies McKay: the McKay correspondence as an equivalence of derived categories, arXiv preprint, arXiv:math/9908027.

[171] Y. Ito, H. Nakajima, McKay correspondence and Hilbert schemes in dimension three, Topology 39 (6) (2000) 1155–1191.

[172] T. Bridgeland, A. King, M. Reid, The McKay correspondence as an equivalence of derived categories, Journal of the American Mathematical Society 14 (3) (2001) 535–554.

[173] A. Craw, An explicit construction of the McKay correspondence for A-Hilb C3, Journal of Algebra 285 (2) (2005) 682–705.

[174] W. Ebeling, A McKay correspondence for the Poincaré series of some finite subgroups of $SL(3, C)$, arXiv preprint, arXiv:1712.07985.

[175] W. Ebeling, Poincaré series and monodromy of the simple and unimodal boundary singularities, arXiv preprint, arXiv:0807.4839.

[176] W. Ebeling, D. Ploog, McKay correspondence for the Poincaré series of Kleinian and Fuchsian singularities, Mathematische Annalen 347 (2010) 689–702.

[177] D. Mumford, Hilbert's fourteenth problem–the finite generation of subrings such as rings of invariants, in: Proceedings of Symposia in pure mathematics, American Mathematical Society, 1976, pp. 431–444.

[178] J. McKernan, Mori dream spaces, Japanese Journal of Mathematics 5 (1) (2010) 127–151.

[179] M. Nagata, On the 14-th problem of Hilbert, American Journal of Mathematics 81 (1959) 766.

[180] S. Mukai, Geometric realization of T-shaped root systems and counterexamples to Hilbert's fourteenth problem, in: Algebraic transformation groups and algebraic varieties: Proceedings of the conference interesting algebraic varieties arising in algebraic transformation group theory held at the Erwin Schrödinger Institute, Vienna, October 22–26, 2001, Springer, 2004, pp. 123–129.

[181] C. V. Johnson, D-Branes, Cambridge University Press, 2003.

[182] P. S. Aspinwall, K3 surfaces and string duality, in: Surveys in differential geometry, Vol. 5, International Press, 1999.

[183] K. Wendland, Consistency of orbifold conformal field theories on K3, Advances in Theoretical and Mathematical Physics 5 (2002) 429–456.

[184] D. Gepner, Z. Qiu, Modular invariant partition functions for parafermionic field theories, Nuclear Physics B 285 (1987) 423–453.

[185] A. Cappelli, C. Itzykson, J. Zuber, The ADE classification of minimal and $A_1^{(1)}$ conformal invariant theories, Communications in Mathematical Physics 113 (1) (1987) 1–26.

[186] P. Di Francesco, P. Mathieu, D. Sénéchal, Conformal field theory, Springer Science & Business Media, 1997.

[187] H. Ooguri, J. L. Petersen, A. Taormina, Modular invariant partition functions for the doubly extended $\mathcal{N} = 4$ superconformal algebras, Nuclear Physics B 368 (3) (1992) 611–624.

[188] T. Gannon, The Cappelli–Itzykson–Zuber ADE classification, Reviews in Mathematical Physics 12 (05) (2000) 739–748.

[189] K. Intriligator, B. Wecht, RG fixed points and flows in SQCD with adjoints, Nuclear Physics B 677 (1–2) (2004) 223–272.

[190] C. Curto, Matrix model superpotentials and ADE singularities, Advances in Theoretical and Mathematical Physics 12 (2) (2008) 353 – 404.

[191] Y.-H. He, Some remarks on the finitude of quiver theories, Mathematics 6 (12) (2018) 291.

[192] D. Gaiotto, E. Witten, S-duality of boundary conditions in N=4 super Yang-Mills theory, Advances in Theoretical and Mathematical Physics 13 (3) (2009) 721–896.

[193] K. Kodaira, On compact analytic surfaces, III, Annals of Mathematics 78 (1963) 1–40.

[194] J. J. Heckman, T. Rudelius, Top down approach to 6d SCFTs, Journal of Physics A: Mathematical and Theoretical 52 (9) (2019) 093001.

Index

Printed by Integrated Books International,
United States of America

Nicolas Scheffer
Jean-Philippe Lang

Substance use and creativity